Corporate Social Responsibility in the Arctic

Corporate Social Responsibility in the Arctic considers the new trends and frontiers of Corporate Social Responsibility (CSR) studies that are shaping the future of global business strategy and ethics.

This book systematically approaches the CSR framework of internal and external factors and their impact on the social responsibility of businesses within the sensitive environment of the Arctic. It presents traditional and contemporary models of CSR through case studies of the eight Arctic nations and explores the debates concerning social responsibility and ethical dilemmas related to social and environmental aspects of business operations, society, and ecosystems. Arruda and Johannsdottir also review approaches for engaging stakeholders in social responsibility, socio-environmental standards, and sustainability, according to frameworks like the UN Sustainable Development Goals, the UN Global Compact, the OECD Guidelines for Multinational Enterprises and the Global Reporting Initiative, but also according to the new CSR strategy in the Arctic based on circular economy, blue economy, smart specialization, knowledge-intensive entrepreneurship, and new parameters of education. Overall, this book examines the ways in which the changing climate and rich natural resources of the Arctic provide unique opportunities and challenges for businesses and societies.

This book will be of great interest to students and scholars of CSR, sustainable business, and business ethics.

Gisele M. Arruda is a Professor in Circumpolar Studies in the field of energy, the Arctic, climate change, environment, and society. She is a member of the US Arctic Committee; an honorary member of the University of Aberdeen; a country director of the UK branch of the International Higher Education Teaching and Learning Association, New York, US; and a Council member of Polar Educators International. She belongs to multidisciplinary research groups in Canada, the US, Iceland, Norway, Greenland, and Alaska researching Arctic governance, sea-ice retreat, Arctic sustainable energy systems, smart specialization, climate change, Arctic societal studies, corporate social responsibility, and multicultural aspects of higher education. She is the author of *Renewable Energy for the Arctic: New Perspectives* and

Sustainable Energy Education in the Arctic: The Role of Higher Education, by Routledge.

Lara Johannsdottir is a Professor in Environment and Natural Resources at the University of Iceland (UIce) and a former academic director of the Environment and Natural Resources graduate programme (2019). Johannsdottir is a board member of the Institute of Business Administration, heads the Environmental and Sustainability Committee of the UIce, and is a former board member of the UIce research fund, and the School of Social Sciences evaluation committee. Johannsdottir was a Fulbright Arctic Initiative Scholar (2018–2019) where she collaborated in research with scholars from Canada, Denmark, Finland, Russia, Sweden, and the United States on Arctic-related affairs. Her research focus is on the business sector, in particular the insurance sector, and corporate social responsibility, corporate sustainability, sustainable business models, responsible investments, and Arctic business affairs.

Routledge Research in Sustainability and Business

Corporate Responsibility and Sustainable Development
Exploring the nexus of private and public interests
Edited by Lez Rayman-Bacchus and Philip R. Walsh

Sustainable Entrepreneurship and Social Innovation
Edited by Katerina Nicolopoulou, Mine Karatas-Ozkan, Frank Janssen and John Jermier

Corporate Social Responsibility, Human Rights and the Law
Stéphanie Bijlmakers

Corporate Sustainability
The Next Steps Towards a Sustainable World
Jan Jaap Bouma and Teun Wolters

Strategic Corporate Social Responsibility in Malaysia
Edited by Rusnah Muhamad and Noor Akma Mohd Salleh

Renewable Energy Enterprises in Emerging Markets
Strategic and Operational Challenges
Cle-Anne Gabriel

Accountability, Philosophy and the Natural Environment
Glen Lehman

Corporate Social Responsibility in the Arctic
The New Frontiers of Business, Management, and Enterprise
Gisele M. Arruda and Lara Johannsdottir

For more information about this series, please visit: www.routledge.com/Routledge-Research-in-Sustainability-and-Business/book-series/RRSB

Corporate Social Responsibility in the Arctic

The New Frontiers of Business, Management, and Enterprise

Gisele M. Arruda and Lara Johannsdottir

Routledge
Taylor & Francis Group
LONDON AND NEW YORK

First published 2022
by Routledge
2 Park Square, Milton Park, Abingdon, Oxon OX14 4RN

and by Routledge
605 Third Avenue, New York, NY 10158

Routledge is an imprint of the Taylor & Francis Group, an informa business

British Library Cataloguing-in-Publication Data
A catalogue record for this book is available from the British Library

Library of Congress Cataloging-in-Publication Data
Names: Arruda, Gisele M., author. | Lara Johannsdottir, author.
Title: Corporate social responsibility in the Arctic : the new frontiers of business, management, and enterprise / Gisele M. Arruda and Lara Johannsdottir.
Description: Abingdon, Oxon ; New York, NY : Routledge, 2022. | Series: Routledge research in sustainability and business | Includes bibliographical references and index.
Identifiers: LCCN 2021007185 (print) | LCCN 2021007186 (ebook)
Subjects: LCSH: Social responsibility of business–Arctic Regions. | Sustainable development–Arctic Regions. | Arctic regions–Economic conditions–21st century.
Classification: LCC HD60.5.A68 A77 2022 (print) | LCC HD60.5.A68 (ebook) | DDC 658.4/0809113–dc23
LC record available at https://lccn.loc.gov/2021007185
LC ebook record available at https://lccn.loc.gov/2021007186

ISBN: 978-0-367-52915-4 (hbk)
ISBN: 978-0-367-52918-5 (pbk)
ISBN: 978-1-003-07981-1 (ebk)

Typeset in Times New Roman
by Newgen Publishing UK

Contents

Figures

Tables

Acknowledgements

This book is the fruit of the authors' passion about the Arctic and their deep love for the peoples and ecosystems of the Arctic. It is the product of an intense desire to make the Arctic a sustainable place where businesses, entrepreneurs, managers, scientists, society, and ecosystems can co-exist and thrive. I would like to thank my dear friend Prof. Lara Johannsdottir for her Icelandic genuine spirit and H.E. Guðmundur Ingi Guðbrandsson, Iceland's Minister for the Environment and Natural Resources, for being such enlightened and inspirational examples that make the sustainability mission possible.

Gisele M Arruda

Abbreviations

AAE	Arctic Adaptation Exchange
ACE	Arctic Economic Council
AECO	Arctic Expedition Cruise Operators
ARENA	Arctic Remote Energy Networks Academy
BITC	Business in the Community
CDP	Carbon Disclosure Project
CEO	Chief Executive Officer
CSP	Corporate Social Performance
CSR	Corporate Social Responsibility
DJSI	Dow Jones Sustainability Indexes
ECOSOC	Economic and Social Council
EEA	European Economic Area
EIA	Environmental Impact Assessment
EITI	Extractive Industries Transparency Initiative
EMAS	EU Eco-Management and Audit Scheme
EPA	Environmental Protection Agency
ESG	Environmental, Social, and Governance factors
EU	European Union
FPIC	Free, Prior, and Informed Consent
GNI	Gross National Income
HRCA	Human Rights Compliance Assessment
IAATO	International Association of Antarctica Tour Operators
ICT	Information and Ccommunications Ttechnologies
IFC	International Finance Corporation
IIRC	International Integrated Reporting Council
ILO	International Labour Organization
IPCC	Intergovernmental Panel on Climate Change
IRENA	International Renewable Energy Agency
ISO	International Organization for Standardization
KPIs	Key Performance Indicators
MCDA	Multi-Criteria Ddecision Analysis
MDGs	Millennium Development Goals
NCA	Network-centric Aapproach

NFRD	Non-financial Reporting Ddirective – or Directive 2014/95/EU
OECD	Organisation for Economic Co-operation and Development
PR	Public Relations
PRI	Principles for Responsible Investment
SAI	Social Accountability International
SASB	Sustainability Accounting Standards Board
SDGs	Sustainable Development Goals
SDWG	Sustainable Development Working Group
SDWG	Sustainable Development Working Group
SEI	Stockholm Environment Institute
SLO	Social License to Operate
SMEs	Small and Medium-sized Enterprises
SRC	Stockholm Resilience Centre
TCFD	Task Force on Climate-related Financial Disclosure
UN PRI	United Nations Principles for Responsible Investments
UN	United Nations
UNDRIP	United Nations Declaration on the Rights of Indigenous Peoples
UNEP FI	United Nations Environment Programme Finance Initiative
UNEP	United Nations Environment Programme
UNFCCC	United Nations Framework Convention on Climate Change
UNGC	United Nations Global Compact
UNGPs	United Nations Principles on Businesses and Human Rights
WBCSD	World Business Council for Sustainable Development
WCDRR	World Conference on Disaster Risk Reduction
WCED	World Commission on Environment and Development
WMU	World Maritime University

1 CSR and the new frontiers of business, management, and enterprise

The new CSR strategy in the Arctic

Learning objectives

- Understand the new dynamics of the North in terms of business, management, and enterprise.
- Understand the main features of the Arctic development process.
- Understand the new CSR strategy in the Arctic and its drivers.

Introduction

The Arctic region is increasingly a focus of economic and political interest. New business opportunities are on the rise and Arctic stakeholders are playing a key role in facilitating and creating favourable conditions for boosting economic activity in the Arctic. Over the last decade, the North experienced an increase in a wide spectrum of business activities, management systems, and new enterprise ventures. Traditional activities such as oil and gas, mining, and shipping are operating alongside the blue economy and bioeconomy, tourism, innovation, and entrepreneurship.

The Arctic region holds extraordinary economic potential, which is being unveiled by the dramatic ecosystemic and climatic transformation underway. This economic potential has inspired a vision of promoting and developing indigenous cultural businesses through local economic activities and policies designed to address specific opportunities, risks, and needs of the region. Business development in the Arctic must be based on in its people's creative potential, entrepreneurship vision, and needs. The main drive is to create innovative development which respects the support capacity of the Arctic region and the planet. It is a vision based on the synergic combination of the Arctic's unique resources and human capital to make its economic activities competitive on a global scale.

Arctic sustainable economic growth based on social responsibility stimulates a durable approach to job creation, increases the use of environmentally efficient technologies, and provides innovative solutions for sociocultural affairs in the region. Coordination and cooperation among the Arctic states, industry, and civil society organizations are key components of this process, and will pave the way for a unique approach in conducting sustainable business in the Arctic. Responsible business practices involve not only respect for human rights, labour regulation, and environmental standards but a deep understanding and respect of multicultural differences and perspectives on development.

Types of discussion and issues explored in Chapter 1

This is an introductory chapter aiming at presenting an overview of the concept and the main components of the new corporate social responsibility (CSR) strategy in the Arctic. It also provides a clear understanding of the new dynamics of the North in terms of business, management, and enterprise. The Arctic has become the main global resource base and a major trade corridor, triggering specific and identified risks for the governance and development process of this new frontier of business, management, and enterprise. It is considered a developing region but this cannot be seen as a conventional development process like that experienced in the South over the last centuries. There are intrinsic and multidisciplinary complexities with amplified effects at local, regional, and global levels. This chapter discusses the relevance of the new dynamics of the North through the lenses of extractive industries, sustainable development, and the well-being of the region's indigenous population, themes that will be extensively amplified in specific forthcoming sections. It also introduces the contemporary CSR trends in the region that will be further debated in subsequent chapters.

Background: Remote versus urban Arctic areas

The Arctic context has rural and urban facets. Rural Arctic communities are under significant pressure due to the environmental change that is currently transforming local economic activity and livelihoods. Younger generations are experiencing modernity and want services, quality of life, opportunities, education, technology, and the whole spectrum of things offered in Arctic urban areas. On the other hand, Arctic urbanization has highlighted the existential necessity of Arctic business development for the creation of scientific, technological, educational, and economic hubs stimulating the emergence and networking of a critical mass able to offer younger generations the fruits of modernity they want to experience and embed in their lives.

This new scenario of socio-environmental aspirations and new technologies in transport, communication, extractive industry, raw material processing, and education have disrupted the traditional economic activities and traditional business models by reorienting business priorities, principles, innovation capacity, and operations in this sensitive region. This dynamic has affected specific aspects of life in the North; environmental sensitivities and daily social needs are naturally forcing businesses to reinvent, redesign, and innovate their strategies. The perception of a sense of remoteness has changed. Indigenous communities are no longer understood solely in geographical terms, and through new media and technology can pursue community-focused goals, exchange information, preserve oral traditions, connect dispersed communities, and communicate and share resources to collectively tackle social issues affecting their well-being and identity (Srinivasan, 2006; Arruda and Krutkowski, 2016). Remote sensing has become a crucial technological tool to monitor climate change and its interactions. It has also helped to develop a better geographical information system (GIS) driving important scientific investigations. With applications in climatology, environmental and social sciences, GIS is used to assess levels of development, land management, natural resources management, and forest biomass, and to map minerals and environmental changes (ESA, 2017).

The availability of energy resources has reawakened government interest in the remote parts of the Arctic. Renewable energy projects have been implemented by governmental partnerships with companies and universities employing the best of the local talents to transform energy systems, but also empowering local communities (Arruda, 2018; Marsik and Wiltse, 2018). This transition to renewable energy, materialized in local projects in several scattered parts of the Arctic, reveals an important trend of integration into the Arctic landscape as well as the interconnectivity of the Arctic social structure. The development of microgrids for the Alaskan and Canadian communities goes beyond being part of a modern trend; it represents the dream of energy access via technological innovation, local know-how, and expertise being applied to society. The integration or interconnectivity provides social interaction and the benefit of creating smart communities that have increased access to communication (Verk et al., 2019), education, training, and online learning.

The Arctic states have shown a clear intention of promoting a sustainability agenda focused on a more sophisticated strategy and balance between economic, social, and ecological dimensions. The intergenerational needs of promoting sustainable economic development in the Arctic requires a more in-depth understanding of the region's business environment, stakeholders, entrepreneurship, governance, and policy incentives in the short and the long run. There also needs to be a clear identification of risks and opportunities for unlocking the economic and human potential of the Arctic for the benefit of the entire region and the world.

The new CSR strategy in the Arctic

Under the retreating permafrost lies a wealth of raw materials like gold, titanium, rare earths, and uranium, among others, which stirs up economic interests from Arctic and non-Arctic nations and an intense resource race as the manifestation of the 21st century technology war. In other words, there is a new race to secure untapped mineral and energy resources for the benefit of state and corporate actors. The new resource race risks a disorganized pattern of development based on demand and supply-chain needs instead of the needs of people. The problem is that this kind of developmental action is fundamentally incompatible with the complex nature of the Arctic and its sustainability principles and SDG targets.

The unbalanced economic policies ignoring the socio-environmental components of sustainability provide a negative example of resources-oriented policies and have negatively influenced the dismantling of environmental policies shaped over decades of political efforts to protect biodiversity and indigenous populations. This phenomenon has created new contexts in which a further threat is the reorientation of the extractive industry to new frontiers of development, i.e. Greenland. The interconnection of South and North supply chains raises this specific risk of a disarticulated and myopic vision of Arctic development based on the same premises as the old style of development and colonialism from previous centuries.

With this perspective in mind, the Nordic Council of Ministers and the Arctic Economic Council elaborated a joint strategy to support economic development in the Arctic through business analysis by mapping fields like entrepreneurship and innovation, public–private partnerships (PPPs) and business cooperation, bioeconomy, and creative and cultural industries. This is a coordinated effort to acknowledge the needs and the necessary measures to promote an overarching framework of what can be termed 'differentiated responsibility' or the new dimension of Arctic responsibility.

The new Arctic CSR vision and strategy relies on a better understanding of the Arctic business environment, its drivers, capabilities, and resources. It also requires knowledge of effective consolidated practices as well as the mechanisms of adherence (Arruda, 2019) of new practices to the United Nations Sustainable Development Goals (SDGs) as objective and global parameters to ensure sustainable development. The business environment and practices of the Arctic should comply with durable, inclusive, and sustainable economic growth (SDG 8); sustainable industrialization and promotion of innovation (SDG 9); sustainable economic activities related to bioeconomy, creative, and cultural industries; business cooperation towards low-carbon energy production and consumption; affordable and clean energy (SDG 7); sustainable cities and communities (SDG 11); and life below water and on land (SDG 14 and 15), among other factors (NCM, 2018, p. 2).

The key driver of the new CSR strategy is the alignment and adherence of mindset and policies to the SDGs (Arruda, 2019). These can be organized into four main categories showing economies and societies as embedded parts of the following aspects:

- Biosphere: The world will fail to achieve the remaining goals unless we achieve the goals related to clean water and sanitation, life below water, life on land and climate action.
- Society: Calls for eradication of poverty, improvement of social justice, peace and good health.

- Economy: Attention towards industry, innovation and infrastructure, reduced inequalities, responsible consumption and production, decent work and economic growth decoupled from environmental degradation.
- Partnerships: Strengthening global solidarity. An integrated approach is crucial for progress across the multiple goals.

(DNV GL, 2019, p. 7)

To shape a sustainable society and a sustainable future, it is fundamental to align knowledge systems, communication, education, policies, and business and management practices with clear sustainable parameters (Arruda, 2019). The 2030 Agenda and the Sustainable Development Goals (SDGs) adopted in 2015 represent these necessary parameters and are the basis for a global plan of action composed of building blocks (Ganapin, 2018) establishing the sustainability parameters for humankind. A global framework of 232 unique indicators has been developed to monitor progress towards the SDG targets; however, work remains to be done in terms of mapping and developing specific indicators because not all the indicators in the global framework are developed in a way to allow monitoring and statistical measurements at a national and local level. The SDGs and their respective indicators represent fundamental guidelines for socio-environmental and economic transformation and measurement, as well as a clear path for a more integrated approach to sustainable development inspiring a new level of Arctic responsibility (see Chapters 3 and 8).

Entrepreneurship and innovation in the Arctic (knowledge-intensive entrepreneurship)

Under the idea of developing a resilient (blue/green/bio) economy in the Arctic, it is important to envisage other drivers of local business development apart from corporations. Entrepreneurship and innovation are also important drivers to be considered from this new responsibility viewpoint. The Merriam-Webster dictionary defines entrepreneurship as "the activity of organizing, managing, and assuming the risks of a business enterprise" (Merriam-Webster, 2019), while the Cambridge Dictionary defines entrepreneurship as "skill in starting new businesses, especially when this involves seeing new opportunities" (Cambridge English Dictionary, 2019).

The concept of entrepreneurship is broad and comprehensive, consisting of several components aligned to the sense of ownership, independence, creativity, innovation, flexibility, problem-solving, impact, responsibility, passion, and vision for the future. The Arctic itself requires innovation and the Arctic business environment is also based on entrepreneurship and innovation, which depend on the development of an entrepreneurial mindset, collaboration among stakeholders and regions, and an open mind to avoid cultural bias against the Arctic region.

High-growth companies, SMEs, and start-ups are part of this new frontier of business, management, and enterprise based on a higher level of personal and collective responsibility. They are perceived not only as positive vectors for job creation, productivity growth, and innovation, but also as an integrative force to promote a wider and more optimized use of industrial and environmental biotechnologies and multicultural inclusiveness. Entrepreneurship is an important stimulus to foster the emergence of industrial and environmental biotechnology as a growing segment of the Arctic bioeconomy (see Chapters 4 and 7). Of particular relevance is the ability of developing regions to participate in the new bioeconomy via the adoption of emerging technologies, an open market access system, flexible enforcement of intellectual property rights, corporate alliances, and effective systems for managing the economic risks and benefits associated with the use of new technologies (Juma and Konde, 2002).

The current and next generation of innovative Arctic entrepreneurs acknowledge the potential of Arctic excellence, multidisciplinary research, training, networks of talents, ideas, and companies, but above all the importance of connecting learning and knowledge to the Arctic environment and society. The young Arctic generation find inspiration from cultural diversity, Arctic creative expertise, and a better understanding of the Arctic business, management, and working environment. They have explored business interfaces, multidisciplinary research, and intensive knowledge entrepreneurship (see Chapter 7). As a result, they are able to promote Arctic Smart Specialization strategies (Foray, 2018) (see Chapter 7), encourage investment and co-management (Arruda, 2018a), and create future domestic capability and interregional competitive advantage by creating an Arctic brand. The Arctic entrepreneurs are an important part of the new Arctic responsibility strategy as they stimulate the development of regions based on new combinations of knowledge and industries. Knowledge is the power of the North.

Further resources: The next chapters

Important themes are analysed in the chapters of this book through the lenses of the new CSR strategy in the Arctic. There is a recognition of the intimacy of relationships between corporations and society, and we emphasize that such relationships must be in the minds of managers as corporations and related groups pursue their respective goals. In Chapter 2, concepts and frameworks are analysed and serve as a crucial parameter for comparisons among the different phases of CSR evolution and the important transition the Arctic region is experiencing currently. Aligned with this evolution, Chapter 3 approaches CSR in a world in transition by presenting the contemporary trends related to climate change impacts in the Arctic, climate change adaptation in the context of ecological and socio-economic changes and in relation to the concept of resilience. Chapter 4 focuses on business practices rather than theory by analysing case studies of the eight Arctic nations and their respective industries. The cases covered ranged from fine art and culture to mining and hydrocarbons, from small newly founded companies to large industry groups and networks, and from cooperatives to government-owned and limited liability companies. Chapter 5 explores the role of business and stakeholders in the new CSR strategy in the Arctic. It addresses what distinguishes contemporary stakeholder theory from the Arctic stakeholder theory and the importance of socio-environmental Arctic business standards and stakeholder dialogue for operating in a responsible manner in the region. The main message of Chapter 6 is about the fundamental role CSR plays in strengthening the relationship between businesses and communities in a sustainable way. Chapter 7 also emphasizes this message, with a focus on the practical solutions CSR can offer in complex environmental contexts and the key environmental pressures and global threats to the 21st-century Arctic. Chapter 8 amplifies the new trends in business practices and sustainable enterprise, addressing the existing CSR frameworks and guidelines for the Arctic, such as the OECD guidelines for multinational enterprises, the UN Global Compact, the UN Declaration on the Rights of Indigenous Peoples (ILO Convention # 169), and the CSR Compass (Nordic Council of Ministers) for Small and Medium Enterprises. Chapter 9 focuses on auditing and reporting tools, as well as conventions, principles, guidelines, and standards that are recognized as fundamental in the Arctic context. Chapter 10 brings the analysis to a conclusion by discussing the future trends of CSR and the crucial role of education in realizing the new CSR strategy in the Arctic through promoting responsible multicultural interactions and forming responsible leaders of the 21st century.

Case study: Arctic innovative entrepreneurs – the next generation

As previously noted, Arctic urban areas have recently received important attention as a source of jobs and opportunities. The Nordic Smart City Network is a collaboration initiative that launched with the participation of 20 Nordic cities in five Nordic countries. These cities share common goals of exploring the Nordic way to create sustainable smart cities based on the premise of maintaining well-being and competitiveness in the global market. This collaborative effort consists in sharing the same vision, values, experiences, best practices, and networks to provide innovative solutions to practical urban issues and encourage the active role of citizens in society. The Nordic network has a core aim of making the cities better for citizens. The Nordic Network of Smart Cities is linked to a joint second thematic project called Nordic Urban Living Labs, which started in 2018 and is financed by Nordic Innovation. The aim of this project is to share best practices for urban development and data gathering to enhance socio-environmental quality as well as attract global investment. See more at https://nscn.eu/Citylabs and https://nscn.eu/.

Considering the negative effects of the COVID-19 pandemic, and with the same spirit of enhancing citizens' quality of life, another relevant case study is Nordic Healthy Cities. In collaboration with the Nordic Smart City Network

(NSCN) that runs until April 2022, it aims to create supportive urban environments to improve health and quality of life. NSCN works in cooperation with private companies to deliver sustainable solutions to address the health challenges brought by increased urbanization. It also intends to mitigate and prevent health challenges, turning the public sector into a driver of innovation to transform urban centres into functional city developments, providing health and quality of life, climate protection initiatives, and sharing knowledge and learning. Please see more at https://nscn.eu/NordicHealthyCities (Nordic Health Cities, 2020).

Questions for discussion

1. What are the main drivers of the new CSR strategy in the Arctic?
2. What is the role of companies and Arctic innovative entrepreneurs in the realization of the CSR strategy in the Arctic?
3. How does the blue economy relate to entrepreneurship?
4. What does this new level of responsibility mean for the Arctic?

Chapter summary

This first chapter introduced the vision of the new CSR strategy in the Arctic. The Arctic with its unique characteristics requires a different level of individual and collective responsibility that does not follow the traditional patterns of South CSR. It must be seen, however, as a new approach to the complexities of an integrated and interdependent global scenario in which people from different cultural backgrounds need to be part of development (Arruda, 2019). The reflection presented in this chapter and the subsequent chapters of this book introduce the idea of a higher level of responsibility the Arctic and non-Arctic world. The vision and lessons applied to the Arctic can inspire a global movement of awareness and citizenship. The young generation need to have a voice and be part of this developmental process that will determine and shape the future of civilization. In the following chapters, the concept and characteristics of this new vision will be expanded and discussed. We hope you enjoy this fascinating learning journey.

References

Arruda, G.M. (2018). *Renewable Energy for the Arctic: New Perspectives*. Abingdon, UK: Routledge.

Arruda, G.M. (2018a). Arctic resource development: A sustainable prosperity project of co-management. In G.M. Arruda (Ed.), *Renewable Energy for the Arctic: New Perspectives* (pp. 106–117). Abingdon, UK: Routledge.

Arruda, G.M. (2019). *Sustainable Energy Education in the Arctic: The Role of Higher Education* (1st ed.). Abingdon, UK: Routledge.

Arruda, G.M. and Krutkowski, S. (2016). Arctic governance, indigenous knowledge, science and technology in times of climate change: Self-realization, recognition, representativeness. *Journal of Enterprising Communities: People and Places in the Global Economy*, 11(4), 514–528.

Cambridge English Dictionary. (2019). 'Entrepreneurship': Definition in the Cambridge English Dictionary. Accessed on 12 December 2020 from https://dictionary.cambridge.org/us/dictionary/english/entrepreneurship

DNV GL. (2019). *Sustainable Blue Economy in the Norwegian Arctic*. Centre for the Ocean and the Arctic. Accessed on 16 December 2020 from www.dnvgl.com/publications/part-1-sustainable-blue-economy-in-the-norwegian-arctic-165023

European Space Agency (ESA). (2017). 'Earth from Space'. Accessed on 30 December 2020 from www.wsa.int/Our_Activities/Observing_the_Earth/Copernicus/Sentinel-2

Foray, D. (2018). Smart specialization strategies as a case of mission-oriented policy: A case study on the emergence of new policy practices. *Industrial and Corporate Change*, 27, 817–832.

Ganapin, D. (2018). The UNSDGs: Providing the building blocks for a more sustainable future. *The Circle*, 2, 3–28.

Juma, C. and Konde, V. (2002). *The New Bioeconomy. Industrial and Environmental Biotechnology in Developing Countries*. UNCTAD/DITC/TED/12. New York: United Nations.

Marsik, T. and Wiltse, N. (2018). A low carbon Arctic energy system? Challenges, opportunities and trends. In G.M. Arruda (Ed.), *Renewable Energy for the Arctic: New Perspectives* (pp. 3–24). Abingdon, UK: Routledge.

Merriam-Webster. (2019). 'Entrepreneurship'. Definition in Merriam-Webster Dictionary. Accessed on 12 December 2020 from www.merriamwebster.com/dictionary/entrepreneurship

Nordic Council of Ministers. (2018). *The Nordic Arctic Business Environment: Entrepreneurship and Innovation*. København, Denmark: Nordic Council of Ministers. Accessed on 20 December 2020 from www.norden.org/en/publication/arctic-business-analysis-entrepreneurship-and-innovation

Nordic Health Cities. (2020). Nordic Health Cities Project. Accessed on 20 December 2020 from https://nscn.eu/NordicHealthyCities

Nordic Innovation. (2021). Smart City Challenge 2021. Accessed on 4 January 2021 from www.nordicinnovation.org/news/nordic-innovation-partners-nordic-smart-city-challenge-2021

Srinivasan, R. (2006). Indigenous, ethnic and cultural articulations of new media. *International Journal of Cultural Studies*, 9(4), 497–518.

Verk, N., Golob, U., and Podnar, K. (2019).. A dynamic review of the emergence of corporate social responsibility communication. *Journal of Business Ethics*, 1–25. doi.org/10.1007/s10551-019-04232-6

2 Historical background

CSR timeline, concepts, frameworks, and theories

Learning objectives

- Recognize the historical development and milestones in the development of CSR.
- Understand underlying concepts, frameworks, and theories of CSR.
- Understand how the Arctic Council's understanding of CSR aligns with underlying concepts, frameworks, and theories of CSR.
- Identify which theories are most relevant in the Arctic business context in view of the Arctic CSR and sustainability framework.
- Employ the underlying concepts, frameworks, and theories of CSR to a real-life case study.

Introduction

The idea that corporations have social responsibility is not new, but the focus in this chapter is on the modern understanding of the corporate social responsibility (CSR) concept, and how it has influenced theory, research, and practice. This understanding is usually traced back to the 1950s (Carroll, 1999, 2008; Latapí et al., 2019), although the discussion about businessmen (now business executives) and their responsibility started earlier (Bowen, 1953; Carroll, 1999). The academic study of CSR requires knowledge of the historical development of the subject and a clear underlying of the timeline, concepts, theories, and frameworks to ensure a clear understanding of contemporary CSR trends.

The Arctic Council's (2013a) understanding and commitment to CSR is emphasized in the Kiruna Arctic Council Ministerial Declaration. The Declaration states that the Ministers:

> welcome the Arctic Council's work on corporate social responsibility and sustainable business, and encourage enterprises operating in the Arctic to respect international guidelines and principles. By this the Arctic Council recognize[s] the central role of business in the development of the Arctic, and decide[s] to increase cooperation and interaction with the business community to advance sustainable development in the Arctic.

The Arctic Council's (n.d.) understanding of the CSR concept is that it entails a voluntary commitment of private entities, which should conduct ethical business operation in interaction with stakeholders and take social and environmental aspects into account in their daily operations. Businesses should furthermore be proactive when protecting the frail circumpolar Arctic environment, and through economic development should improve the quality of life for relevant stakeholders, including employees, their families, local communities, and society at large. Businesses should respect cultural differences, build the skills of their employees and local communities, and protect the environment for the benefit of future generations. The Arctic Council does not stress the importance of developing its own guidelines and principles, but refers to international work of relevance. This includes the OECD Guidelines for Multinational Enterprises, the UN Global Compact, the Global Reporting Initiative, and the CSR Compass for Small and Medium Enterprises (Arctic Council, 2013b); these are all discussed in other chapters of this book, including Chapter 9.

Types of historical milestones, concepts, frameworks, and theories discussed in Chapter 2

The chapter first introduces the background of CSR by discussing relevant historical milestones in the CSR, sustainability, and Indigenous Peoples timeline, presented in Tables 2.1 to 2.6. Each table, except for Table 2.1, covers milestones over a period of ten years. The periods are classified in the following manner: 1953–1969: Initial modern phase of CSR; 1970–1979: Early modern CSR phase and institutional debate; 1980–1989: Period of early operationalization of CSR; 1990–1999: Period of later operationalization of CSR; 2000–2009: Period of mature, strategic, and institutionalized approach to CSR; and 2010–2019: Period of reporting, disclosure, and results. Following the discussion of historical CSR milestones presented in the CSR timeline is a discussion about CSR concepts and other concepts interrelated and overlapping with the CSR concept. Subsequently, three conceptual CSR frameworks, and an Arctic-specific CSR and sustainability framework are presented. CSR theories are then introduced, categorized as instrumental theories, political theories, integrative theories, and ethical theories. The case studied in Chapter 2 is Albatros Expeditions, a Scandinavian company which provides year-round small ship expeditions in the Arctic and Antarctic.

Background: Historical milestones in the CSR, sustainability, and Indigenous Peoples timeline

The development of the modern CSR concept is very much aligned with academic publications, establishment of institutions, and international agreements. Some of the critical milestones influencing CSR are listed in Tables 2.1 to 2.6, beginning in 1953. The timeline overlaps with important

Table 2.1 Historical milestones in the CSR timeline, 1953–1969

1953–1969: Initial modern phase of CSR	
1953	Publication of Bowen's landmark book *Social Responsibilities of the Businessman* and his iron law of responsibility.
1957	Indigenous and Tribal Populations Convention (No. 107) of the International Labour Organization (ILO).
1960	Davis' social responsibility definition.
1960	Frederick's social responsibility definition.
1962	Publication of Carson's book *The Silent Spring.*
1963	Publication of McGuire's book *Business and Society*.
1966	Davis and Blomstrom's social responsibility definition.
1967	Publication of Walton's book *Corporate Social Responsibilities.*

sustainability milestones, which are included since they influence what corporate responsibility entails. Given the scope of this book, events related to the rights of Indigenous Peoples are also included. Where applicable, given the scope of the chapter, milestones covered in the tables are discussed in more detail, but in other cases they are addressed in other chapters of the book where most relevant. The timeline only provides a rough idea about important events, academic publications, establishment of institutions, and international agreements that are important in the CSR timeline since an all-inclusive list would be far too long to include in the book. Therefore, events related to some of the international conventions, industrial disasters, oil spills, and pollution are not included; instead, the establishment of Arctic-related institutions, conventions, and other important milestone are highlighted.

The initial phase of CSR was characterized by landmark definitions and arguments about the role of businesses in society (Bowen, 1953; Davis, 1960; Frederick, 1960; Davis and Blomstrom, 1966; McGuire, 1963), thereby setting the foundation for an academic discussion about the concept; see further discussion about concepts below. The Indigenous and Tribal Populations Convention (No. 107) of the International Labour Organization (ILO) marked a milestone in the rights of Indigenous Peoples: this international instrument was adopted to protect Indigenous Peoples from oppression and discrimination (International Labour Organization, 1957). In addition, Rachel Carson's book, *The Silent Spring*, was very influential in the environmental movement by documenting the harmful impacts of pesticide use on the natural environment (Carlson, 1962).

The early modern CSR phase and institutional debate starts with Milton Friedman's publication, *The Social Responsibility of Business is to Increase its Profits*, where he defines CSR in the following manner:

> there is one and only one social responsibility of business — to use its resources and engage in activities designed to increase its profits so long

Table 2.2 Historical milestones in the CSR timeline, 1970–1979

1970–1979: Early modern CSR phase and institutional debate	
1970	Milton Friedman's publication *The Social Responsibility of Business is to Increase its Profits.*
1970	Establishment of the US Environmental Protection Agency (EPA).
1970	The first Earth Day is held in the United States.
1971	Johnson's definition of a socially responsible firm.
1971	OECD claims those causing pollution should pay, thereby establishing the Polluter Pays Principle.
1971	Committee for Economic Development (CED) publishes *Social Responsibilities of Business Corporations.*
1971	Publication of Steiner's first edition of *Business and Society.*
1972	UN Stockholm Conference on the Human Environment. Establishment of many national environmental protection agencies and the UN Environment Programme (UNEP).
1972	Limits to growth published by the Club of Rome.
1978	OECD Directorate of the Environment relaunches research on the linkage between the environment and the economy.
1975	Sethi's discussion of corporate social performance.
1979	Carroll's four-part definition of CSR.

> as it stays within the rules of the game, which is to say, engages in open and free competition without deception or fraud.
>
> (Friedman, 1970, p. 32)

This doctrine is the foundation for *shareholder theory*, which became very influential but at the same time was very much contested until *stakeholder theory* emerged in 1984 (Friedman, 1970; Freeman, 1984). The decade is also characterized by the establishment of the US Environmental Protection Agency (EPA) in 1970, the United Nations Environment Programme (UNEP) in 1972, many national environmental protection agencies, and environmental conventions. The institutional debate on pollution, limits to growth, and the linkage between the environment and the economy began (OECD, n.d.; Club of Rome, 1972), as well as academic discussion and definitions of the role of businesses in society and what being a socially responsible firm entailed (Johnson, 1971; Steiner, 1971; Carroll, 1979;), and the concept of corporate social performance (Sethi, 1975). The first Earth Day, held in the United States on 22 April 1979, is said to mark the birth of the modern environmental movement (Earth Day, n.d.).

The period of early operationalization of CSR provides the notion of corporate obligations (Jones, 1980), the development of the SA8000 auditable certification standard (1989) aiming at ensuring socially acceptable practices in the workplace, the stakeholder approach (Freeman, 1984), and the business case for CSR (Cochran and Wood, 1984). More alarming are the predictions in 1985 by the World Meteorological Society, UNEP, and

Table 2.3 Historical milestones in the CSR timeline, 1980–1989

1980–1989: Period of early operationalization of CSR	
1980	Jones' CSR definition including the notion of corporate obligations.
1982	Business in the Community, a charity of Charles, Prince of Wales, is founded. Focuses on responsible business, CSR, and corporate responsibility.
1982	Economic and Social Council (ECOSOC) establishes the Working Group on Indigenous Populations.
1983	Grameen Bank influences the understanding of the role of microcredit for development.
1984	R. Edward Freeman publishes *Strategic Management: A Stakeholder Approach.*
1984	Cochran and Wood's empirical study on the relationship between CSR and financial performance.
1985	The World Meteorological Society, UNEP, and the International Council of Scientific Unions predict global warming.
1986	Chernobyl nuclear power plant accident north of Ukraine.
1987	*Our Common Future* by the WCED (Brundtland report) is published.
1987	Montreal Protocol on Substances that Deplete the Ozone Layer adopted.
1988	Intergovernmental Panel on Climate Change (IPCC) is established.
1989	Exxon Valdez runs the ground. Spills 11 million gallons of oil in Prince William Sound, Alaska.
1989	Social Accountability International launches the SA8000 auditable certification standard.
1989	Indigenous and Tribal Peoples Convention, an International Labour Organization Convention.

the International Council of Scientific Unions of global warming, leading to the establishment of the Intergovernmental Panel on Climate Change (IPCC) in 1988. Furthermore, a milestone was marked with the publication of *Our Common Future* by the World Commission on Environment and Development (WCED) in 1987, which popularized the term 'sustainable development'. The Grameen Bank also influenced a new understanding of the role of finance (microcredit) in development by providing finance to the poorest population in Bangladesh (Khandker et al., 1994). The Montreal Protocol on Substances that Deplete the Ozone Layer was furthermore adopted (United Nations Environment Programme, n.d.), but major industrial disasters and oil spills also occurred, for instance the Chernobyl nuclear power plant accident in 1986 (International Nuclear Safety Advisory Group, 1992), and the Exxon Valdez oil spill in Alaska in 1989 (Skinner and Reilly, 1989). Two milestones of significance to the Indigenous Peoples occurred, namely the establishment of the Working Group on Indigenous Populations by the Economic and Social Council (ECOSOC) in 1982 (Office of the High Commissioner for Human Rights, n.d.) and the International Labour Organization Convention on Indigenous and Tribal Peoples, also known as ILO Convention 169, or C169.

Table 2.4 Historical milestones in the CSR timeline, 1990–1999

1990–1999: Period of later operationalization of CSR	
1991	Carroll's four categories or components of CSR presented as the pyramid model.
1992	Rio Earth Summit and the Rio Declaration.
1992	The Business Council for Sustainable Development publishes *Changing Course*.
1992	Businesses for Social Responsibility is founded.
1993	Inaugural meeting of the UN Commission on Sustainable Development.
1995	World Business Council for Sustainable Development founded.
1995	World Trade Organization (WTO) is founded.
1995	Swanson's corporate social performance (CSP) model.
1996	Arctic Council is founded.
1996	ISO 14001 adopts a voluntary standard for corporate environmental management.
1996	Burke and Logsdon's strategic dimensions of value creation.
1997	Kyoto Protocol adopted. Effective from 2005 with a 2008–2012 commitment period.
1997	Epstein's contribution to business ethics.
1999	Launch of the Dow Jones Sustainability Indices (DJSI).

The period of later operationalization of CSR includes a strong institutional focus such as the 1992 United Nations Conference on Environment and Development, also known as the Rio Summit or the Earth Summit, inspiring business interest in the issues discussed and solutions proposed. The Rio Earth Summit resulted, for instance, in an agreement on the United Nations Framework Convention on Climate Change and the United Nations UN Rio Declaration on Environment and Development, or the Rio Declaration, which set forth principles recognizing the importance of preserving the natural environment and offering guidelines for doing so (United Nations, n.d.c). The Kyoto Protocol was adopted in 1997 and became effective in 2005, with a commitment period from 2008 until the end of 2012 (United Nations, n.d.d). The inaugural meeting of the United Nations Commission on Sustainable Development was held in 1993 (Union of International Association, n.d.). Furthermore, the World Trade Organization (WTO) was founded in 1995 with a formal recognition of the link between trade, environment, and development. In 1996, the Arctic Council was founded, a high-level intergovernmental forum addressing Arctic governmental and indigenous people's affairs.

Business and trade-related institutions were also established. These and academic publications inspired business interest in sustainable development, including the Business Council for Sustainable Development publication *Changing Course* (Schmidheiny and BCSD, 1992), establishment of the Businesses for Social Responsibility, a network of companies that want to build a sustainable world in 1992 (BRS, n.d.) and the World Business Council for Sustainable Development in 1995. Furthermore, in 1999, the Dow Jones

Table 2.5 Historical milestones in the CSR timeline, 2000–2009

2000–2009: Period of mature, strategic, and institutionalized approach to CSR	
2000	The UN Millennium Development Goals launched. Set to be achieved by 2015.
2000	The United Nations Global Compact is founded.
2000	Carbon Disclosure Project (CDP) is founded.
2001	Wood and Logsdon's theorizing of business citizenship.
2001	Lantos' boundaries of strategic corporate social responsibility.
2002	Global Reporting Initiative (GRI) releases reporting guidelines.
2002	EU resolution concerning CSR and business contribution to sustainable development.
2005	Chandler and Werther's understanding of CSR from a strategic perspective.
2005	The Kyoto Protocol enters into force.
2005	Publication of A *European Roadmap for Businesses: Towards a Sustainable and Competitive Enterprise*.
2006	Porter and Kramer's publication introducing the concept of shared value.
2006	Principles for Responsible Investment founded.
2007	Husted and Allen's strategic approach to CSR for value creation purpose.
2007	UN Declaration on the Rights of Indigenous Peoples (UNDRIP) adopted.
2007	Climate change awareness of the public increases.
2008	Heslin and Ochoa's seven principles guiding strategic CSR and value creation.
2008	AccountAbility introduces the first of its AA1000 Series of principles-based reporting standards.
2009	Scientists introduce the concept of "planetary boundaries".

Sustainability Indices (DJSI) was launched, offering guidance to investors that look for profitable companies that follow sustainable development principles (Corporate Sustainability Assessment, n.d.), and the international standard-setting body, the International Organization for Standardization (ISO), adopted an international voluntary standard on corporate environmental management (ISO, 1996). Academic scholars also influenced the debate in this decade, including Carroll in 1991 with his four categories or components of CSR presented as the pyramid model; Swanson's corporate social performance (CSP) model (1995); Burke and Logsdon's (1996) strategic dimensions of value creation, namely centrality, specificity, proactivity, voluntarism, and visibility; and Epstein's (1987) contribution to business ethics.

The first decade of the 21st century was a period of mature, strategic, and institutionalized approaches to CSR and many widely known and relevant initiatives were launched that remain in use today. The UN Millennium Development Goals (MDGs) were launched in 2000, set to be achieved by 2015, with ultimate goals including ending extreme poverty and hunger, achieving universal primary education, promoting gender equality and empowering women, reducing child mortality, and combatting diseases such as HIV and malaria (United Nations, n.d.b). In 2005, the Kyoto Protocol took effect as a legally binding agreement (UNFCCC, n.d.), with its first commitment period from 2008–2012. The United Nations Global Compact (n.d.) was founded in 2000, a non-binding UN pact aiming to encourage businesses worldwide to adopt socially responsible and sustainable policies and practices and report on their implementation. The Carbon Disclosure Project (CDP) was also founded in 2000, providing a global disclosure system for investors, companies, cities, states, and regions to help them manage their environmental impacts (CDP, 2020). In 2002, the Global Reporting Initiative (GRI) released global reporting guidelines on comprehensive sustainability reporting (GRI, n.d). Furthermore, the United Nations-supported Principles for Responsible Investment (UN PRI) were launched, with the aim of understanding investment implications of environmental, social, and governance (ESG) factors (PRI, n.d.). AccountAbility, which is a global consulting and standard company, introduced the first of its AA1000 series of principles-based reporting standards in 2008.

The European Commission also influenced CSR debates during this period, with resolutions concerning CSR and business contribution to sustainable development in 2002 and 2003, a *European Roadmap for Businesses towards a Sustainable and Competitive Enterprise* in 2005, and a communication from the European Commission (2006) to the European Parliament. The public's awareness of climate change increased during this decade, resulting, for instance, from the documentary, *An Inconvenient Truth* by Al Gore Jr. and the alarming reports by the Intergovernmental Panel on Climate Change (IPCC); Al Gore Jr. and the IPCC shared the Nobel Peace Prize in 2007 for their work. New and improved definitions and concepts emerged, including from Wood and Logsdon (2001), Lantos (2001), Werther and Chandler (2005), Porter and Kramer (2006), Husted and Allen (2007), and Heslin and Ochoa (2008). Towards the end of the decade, the concept "planetary boundaries" was introduced, entailing nine boundaries of the Earth and its systems, relating for instance to ocean acidification, climate change, and freshwater use (Stockholm Resilience Centre, 2009). It is also important to mention the UN Declaration on the Rights of Indigenous Peoples (UNDRIP), a non-legally binding resolution (A/61/295), adopted by the United Nations General Assembly in 2007.

The 2010–2019 period is characterized by reporting, disclosure, and a focus on implementation and results. This includes the launch of the ISO 26000 Guidance on Social Responsibility in 2010 (Latapí et al., 2019), the establishment of the International Integrated Reporting Council in 2010, promoting communication about value creation as a step in the evolution of corporate reporting, the establishment of the UN Guiding Principles on Business and Human Rights in 2011, and the Sustainability Accounting Standards, founded to develop and disseminate sustainability accounting standards, in 2011. Furthermore, the OECD Guidelines for Multinational Enterprises, recommended to multinational enterprises on responsible business conduct,

Table 2.6 Historical milestones in the CSR timeline, 2010–2019

2010–2019: Period of reporting, disclosure, and results	
2010	ISO 26000 Guidance on social responsibility is launched.
2010	UN Guiding Principles on Business and Human Rights established.
2010	International Integrated Reporting Council founded.
2011	EU definition and strategy for CSR (2011–2014).
2011	OECD Guidelines for Multinational Enterprises.
2011	Sustainability Accounting Standards Board founded.
2011	Porter and Kramer's concept of shared value.
2012	Rio +20 conference aiming at reconciling economic and environmental goals of the global community.
2012	Doha Amendment to the Kyoto Protocol, commitment period 2013–2020.
2014	The EU Directive 2014/95/EU on disclosure of non-financial and diversity information by large companies.
2014	Arctic Economic Council inaugural meeting.
2015	The Paris Agreement is adopted.
2015	UN Sustainable Development Goals adopted.
2015	World Economic Forum's (WEF) Global Agenda Council on the Arctic publishes the Arctic Investment Protocol
2016	Chandler's updated definition on strategic CSR and value creation.
2017	Nasdaq Nordic and the Baltics exchanges announce voluntary support program on environmental, social, and governance (ESG) disclosure.
2019	Nasdaq introduces an updated global version of the *ESG Reporting Guide* for public and private companies.

were published, and the EU Directive 2014/95/EU, laying down the rules on disclosure of non-financial and diversity information by large companies, was accepted but came into effect in 2018 (Latapí et al., 2019). Consequently, the Nasdaq introduced a voluntary support programme on environmental, social, and governance (ESG) disclosure for the Nordic and the Baltic countries in 2017, and an improved global version in 2019.

In the regional and global arena, the European Union (2011) published a definition and strategy for CSR (2011–2014), the Doha Amendment to the Kyoto Protocol was made in 2012 (second commitment period), and the Paris Agreement, entered into force 2016 with a global goal of limiting global warming to below 2°C.

The Rio +20 (2012) conference should also be mentioned as an important milestone, given its reconciliation between economic and environmental goals of the global community (United Nations, n.d.c). In addition, Porter and Kramer (2011) presented their idea on shared value and Chandler (2016) updated his definition on strategic CSR and value creation (Latapí et al., 2019). From the Arctic perspective, the inaugural meeting of the Arctic Economic Council, an independent organization facilitating Arctic business-to-business activities, was held in 2014, and the World Economic Forum's (WEF) Global Agenda Council on the Arctic published the Arctic Investment Protocol in 2015.

CSR definitions and concepts

Bowen's 1953 publication, *Social Responsibilities of the Businessman*, is often said to mark the modern CSR era, in which he referred to the "the obligations of businessmen to pursue those policies, to make those decisions, or to follow those lines of action which are desirable in terms of the objectives and values of our society" (p. 6). This was then followed by a definition from Davis (1960) where he argued that decisions and actions of businessmen entailed obligations beyond direct economic interests. His so-called *Iron Law of Responsibility* noted the responsibility of businessmen ought to be "commensurate with their social power" (Carroll, 1999, p. 71). Another scholar influencing early definitions of social responsibility was Frederick (1960, p. 60), who argued that the concept entailed "that businessmen should oversee the operation of an economic system that fulfils the expectations of the public". Furthermore, "production and distribution should enhance total socio-economic welfare ... and not simply for the narrowly circumscribed interests of private persons and firms" (Frederick, 1960, p. 60). The view of corporations, or businessmen, obligations extending beyond economic, legal, and/or technical obligations and interests, was also expressed by McGuire (1963) and Davis and Bloomstrom (1966), who discussed implications for the "whole system" (p. 12). The top management responsibility was also highlighted by Walton (1967, p. 18), who stated: "In short, the new concept of social responsibility recognizes the intimacy of the relationships between the corporation and society and realizes that such relationships must be kept in mind by top managers as the corporation and the related groups pursue their respective goals".

In 1971, the importance of stakeholders came into view when Johnson claimed: "A socially responsible firm is one whose managerial staff balances a multiplicity of interests. Instead of striving only for larger profits for its stockholders, a responsible enterprise also takes into account employees, suppliers, dealers, local communities, and the nation" (p. 50). The Committee for Economic Development (CED) offered a broader understanding of the business role in the society when they claimed that: "Inasmuch as business exists to serve society, its future will depend on the quality of management's response to the changing expectations of the public" (p. 16). Sethi (1975) added to the discussion about the behaviour of corporations by discussing it in the context of "social obligation addressing legal constraints and market forces", "social responsibility going beyond social obligations", and "social responsiveness", in terms of adaptation of corporate actions to the needs of the society (Carroll, 1999). Jones' (1980) contribution was to highlight the obligation towards other groups than stockholders, claiming responsibly beyond legal requirements and union contracts, also seeing CSR as a decision-making process.

Carroll made an important contribution towards defining CSR, first by discussing the "economic, legal, ethical, and discretionary expectations that society has of organizations at a given point in time" (Carroll, 1979, p. 500). This was then presented in the Pyramid of CSR (Carroll, 1991) where companies' role as good corporate citizens was also highlighted in relation to companies' philanthropic role, replacing the discretionary component of the earlier publication. In a 2015 publication, Carroll discussed how various other concepts – business ethics, corporate citizenship, corporate sustainability, stakeholder engagement, and management and shared value – interrelate, overlap, and are incorporated in the CSR concept (Latapí et al., 2019).

Other valuable contributions include those by Wood (1991), who created a corporate social performance model grounded in CSR principles; Burke and Longdon (1996), who defined five dimensions of strategic CSR; Lantos (2001), who discussed a social contract between businesses and society, which may become strategic if companies use these emphasis in their plans for generating profit; and Marrewijk (2003), who from a strategic CSR perspective discussed it as a "response to the new roles and responsibilities of each sector of society" (Latapí et al., 2019, p. 17). In addition, Porter and Kramer (2006, 2011) put forward the concept of the shared value of businesses and society, stating that it should replace the CSR concept. Huster and Allen (2007) tied value creation to social demands, while Heslin and Ochoa (2008) explained seven common principles of strategic CSR. These principles are to 1) cultivate the needed talent, 2) develop new markets, 3) protect labour welfare, 4) reduce the environmental footprint, 5) profit from by-products, 6) involve customers, and 7) greening the supply chain (Latapí et al., 2019). According to Trapp (2012), corporations should reflect in their activities their concerns about social and global issues, even though they do not directly relate to their core business.

Werther and Chandler's (2005) noticeable contribution includes their recognition of a shift from social responsibility as a minimal commitment to it becoming an aspect of strategic importance that may result in competitive advantage. In 2013, they viewed strategic CSR as central to companies' decision-making, their daily core operations, and aligning with stakeholder interests. In 2016, Chandler, building on previous publications, embraced five major components of strategic CSR: 1) companies' culture and planning process, 2) companies actions and core operation, 3) stakeholder needs, 4) mid and long-term planning and management processes, and 5) the optimization of value creation.

In 2002, the European Commission presented a Green Paper on CSR defining it as "a concept whereby companies integrate social and environmental concerns in their business operations and in their interaction with their stakeholders on a voluntary basis" (p. 3), stating a consensus on its main features including going beyond regulatory requirements voluntarily, a long-term perspective, and ties to sustainable development and companies' core business (European Commission, 2002). In a CSR strategy from 2011, the aim was to align both European and global approaches to CSR, including a focus on principles, guidelines, and standards, such as reporting standards.

This strategy also included the following definition of CSR "the responsibility of enterprises for their impacts on society" but it also entailed that companies should have processes in place that focus on human rights and environmental and ethical issues, as well as a core business and stakeholder focus (European Commission, 2011). In a 2019 document from the EU Commission, the ties between CSR and responsible business conduct, which the Commission uses interchangeably, human rights, and sustainability are strengthened, as well as the focus on the EU's role in the following areas: "(i) helping firms across the industry or in individual sectors to work together on best solutions, (ii) providing incentives for the uptake of Corporate Social Responsibility or Responsible Business Conduct, including by setting benchmarks and other requirements, (iii) raising awareness and (iv) providing necessary training" (European Commission, 2019). The EU CSR focal points align very much with those in the ISO 26000 guidelines where social responsibility is defined as:

> [T]he responsibility of an organization for the impacts of its decisions and activities on society and the environment, through transparent and ethical behaviour that: contributes to sustainable development, including health and the welfare of society; takes into account the expectations of stakeholders; is in compliance with applicable law and consistent with international norms of behaviour; and is integrated throughout the organization and practices in its relationships.
>
> (ISO, n.d.)

Other concepts interrelated to, and overlapping with, the CSR concept include accountability, business ethics, corporate citizenship, environmental, sustainability, integrated or CSR reporting, stakeholder theory, sustainable development and business/corporate sustainability, social license to operate, and shared value. This list is not all-inclusive but provides some indication of the nature of the discussion. Some scholars (e.g. Carroll, 2015) claim that such concepts are incorporated into the contemporary understanding of CSR. This is, however, not a universal truth as others may understand CSR as one of the underlying concepts of the corporate sustainability and moral philosophy disciplines. According to this view, other underlying concepts are sustainable development rooted in the disciplines of economics, ecology, and social justice; stakeholder theory originating in strategic management; and corporate accountability theory founded in business law (Wilson, 2003). For simplification, the same understanding will be applied to the concepts of CSR and corporate sustainability in this book given that they are interrelated and overlapping.

CSR frameworks or models

Three commonly known CSR frameworks will be introduced, together with an Arctic-specific one. The first CSR framework is the pyramid model of CSR, introduced by Carroll in 1991. Responsibilities are represented as hierarchal, starting with economic responsibility where business profitability is seen as the foundation for other responsibilities. Next are legal responsibilities, where companies obey the law. This is followed by ethical responsibilities, where companies should "do what is right, just, and fair. Avoid harm" (p. 42). Finally, as part of companies' philanthropic responsibilities is their corporate citizenship where their resources contribute to the well-being of society. In a later revision of the pyramid (Carroll, 2016; see Figure 2.1), it is emphasized that it is a stakeholder framework which should be viewed as integrated, and that "ethical motivations and issues cut through and permeate all four of the CSR categories and thus assume a vital role in the totality of CSR" (p. 5).

The second CSR framework (see Figure 2.2) shows four circles. In this case CSR is not presented in a hierarchal order; instead, responsibilities are interconnected and interrelated. Scholars are free to decide how many circles to include in the framework (usually three to four) and on the main elements.

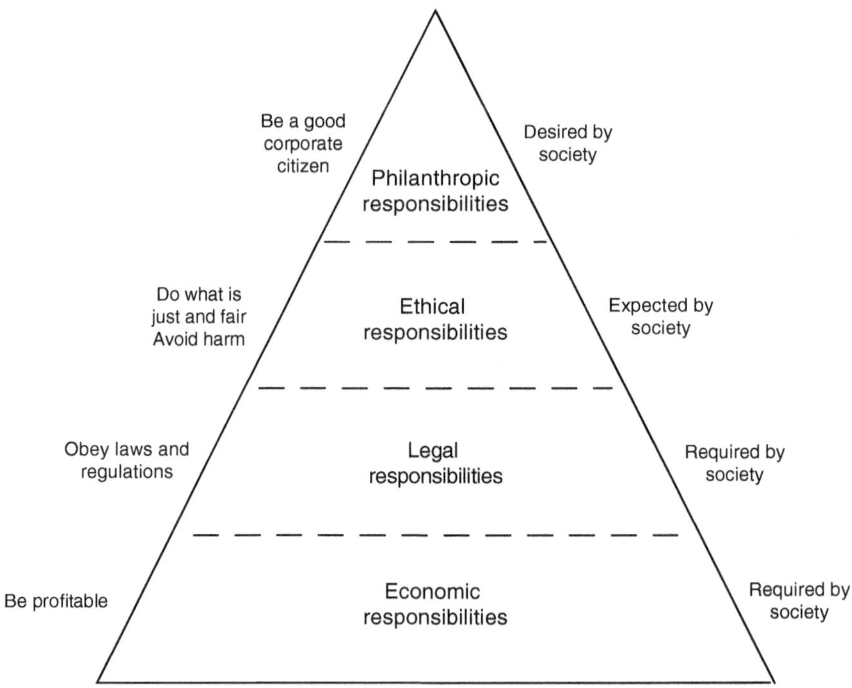

Figure 2.1 Carroll's pyramid of CSR

Carroll's pyramid of CSR emphasizes different types of responsibility that corporations should fulfil. These include economic, legal, ethical, and philanthropic responsibility. These responsibilities should be viewed as integrated rather than viewed in isolation.

Source: Carroll (2016, p. 5)

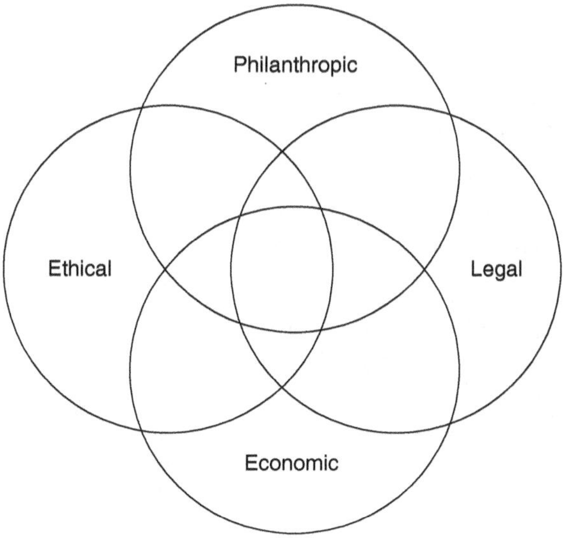

Figure 2.2 Intersecting circles of CSR

The figure emphasizes four types of responsibility that corporations should fulfil. These include economic, legal, ethical, and philanthropic responsibility. These responsibilities are not presented in a hierarchal order; instead, responsibilities are interconnected and interrelated.

Source: Geva (2008, p. 5)

For instance, Jones (1983) included the elements of the economy, culture, policy, and society, Schwartz and Carroll (2003) incorporated economic, legal, and ethical aspects, and Geva (2008) built in economic, legal, ethical, and philanthropic aspects, as demonstrated in Figure 2.2.

The third CSR framework (see Figure 2.3) consists of concentric circles of CSR (Committee for Economic Development, 1971; Geva, 2008). Originally the framework included three circles: an inner circle of economic responsibility such as products, economic growth, and employment; an intermediate circle taking into account social values and main concerns, including environmental responsibility, employee relations, and customers' expectations; and an outer circle where companies should actively improve the social environment, for instance by addressing poverty (Committee for Economic Development, 1971; Carroll, 1999). This type of framework demonstrates a relationship in a system of inclusion, as elements of inner circles are included in the outer circles, but not vice versa. This means that economic responsibilities also include legal, ethical, and philanthropic responsibilities (Geva, 2008).

The fourth CSR framework is Arctic specific and focuses more on sustainability-related risks and Arctic economies. It was originally developed by a group of 2018–2019 Fulbright Arctic Initiative Scholars (Vammen

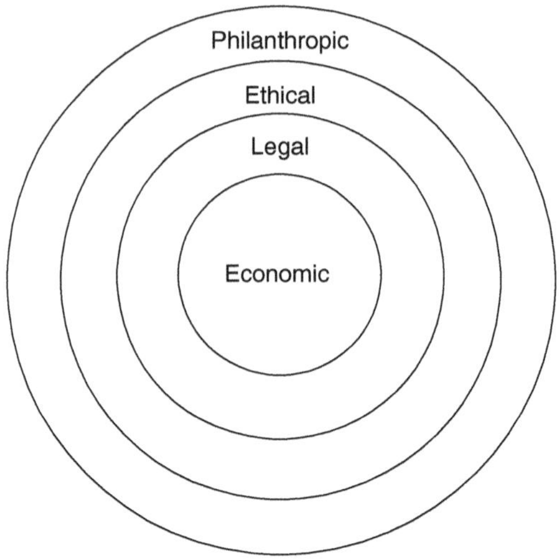

Figure 2.3 Concentric circles of CSR

The concentric circles of CSR shows a relationship in a system of inclusion, as elements of inner circles are included in the outer circles, but not vice versa. The circles from the innermost circle to the outermost circle are economic, legal, ethical, and philanthropical.

Source: Committee for Economic Development (1971); Geva (2008, p. 5)

Larsen et al., 2019), and presented in a more simplified version in a policy brief by the same group of scholars in 2019. The version introduced here has been modified so that it aligns with sustainability, CSR and risk emphasis, and aspects of Arctic business operations and local concerns. The framework should be considered as holistic and integrated. Economic responsibilities are in the centre, ranging from subsistence to market-type activities, on a local and global scale. The legal aspect of doing business in the region is incorporated in the Arctic-specific institutional context, such as the Arctic Council and specific conventions protecting the rights of indigenous peoples. This level is then surrounded by ethical actions. The ethical level, just as with the legal and economic level, needs to take into account Arctic-specific social, cultural, and environmental considerations, which should be integrated with the core activities and daily operation of businesses operating in the Arctic. While the philanthropic aspect is included in the framework, it is considered less important than the risk aspect of Arctic operations when economic opportunities are considered or carried out. The risk assessment should be considered on a systemic level, determining what systems are at risk, in terms of risk assessment, management, and communication (Johannsdottir and Cook, 2019). These systems can for instance be financial systems, environmental systems, and/or social systems.

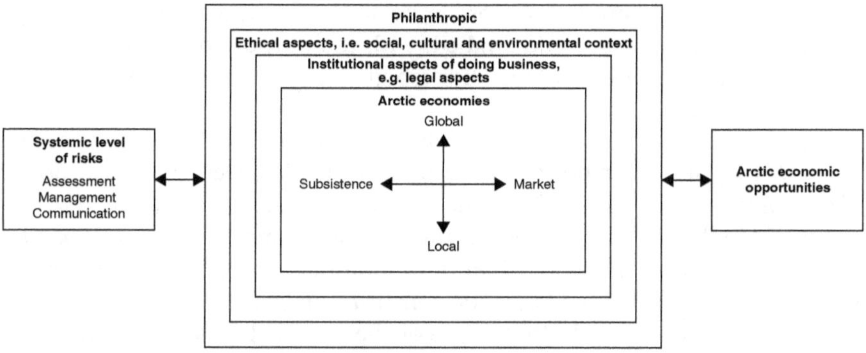

Figure 2.4 Arctic CSR and sustainability framework

The Arctic CSR and sustainability framework has several key elements. It shows the systemic level of risks, where assessment, management, and communication, should be emphasized. It also shows Arctic economic opportunities. Both the risk aspects and the opportunities influence the Arctic economies, which include subsistence, mixed, and market economies, with activities ranging from local to global. These different types of economies are surrounded by the institutional aspect of doing business in the region, ethical aspects, including, social, cultural, and environmental context, and finally the philanthropic aspects.

Source: Johannsdottir and Cook (2019); Vammen Larsen et al. (2019, p. 688)

CSR theories

There are many CSR theories, which may differ in the context and classification. In mapping the territory in 2004, Garriga and Melé classified the theories into four groups, i.e. instrumental theories, political theories, integrative theories, and ethical theories, which will be used as a classification for the following discussion.

Instrumental theories regard the role of corporations to serve as value-creation instruments where social activities are a means to the end of attaining economic results (Garriga and Melé, 2004). Fundamentally, the focus is on the economic aspect of companies' actions in the sense that social activities are accepted only if they result in economic benefits for the business. Friedman's view (1970) exemplifies these types of theories and views, which was for a long time very popular among company owners and managers. If the interests of company stakeholders aligns with companies' motivation for maximizing shareholder value, they should be taken into account; this view is evident in the *agency theory* (Ross, 1973; Jensen and Meckling, 1976). The downside of this theoretical point of view is the focus on short-term profits (Garriga and Melé, 2004). *Enlightened value maximization* involves taking more of a strategic approach to value creation (Jensen, 2000). In order to achieve competitive advantage, the allocation of resources matters so as to achieve long-term social aims (Garriga and Melé, 2004). The business case for CSR, or the ties between corporate social and financial performance, also has been widely discussed and studied, with the purpose of finding out if it pays off for companies to be socially responsible (Margolis and Walsh 2003; Orlitzky et al., 2003).

Political theories centre on the political strength of corporations in society, and how they use their power in the political field. Given the social power of corporations, they should assume social responsibilities and rights. This framing includes three other important theories, namely *Corporate constitutionalism*, *Integrative social contract theory*, and *Corporate Citizenship theory* (Garriga and Melé, 2004). Corporate constitutionalism theory focuses on the business power in society, both in terms of the so-called social power equation principle which states that "social responsibilities of businessmen arise from the amount of social power that they have" (Davis, 1967, p. 48; Davis, 1960) and the iron law of responsibility, which claims that "negative consequences of the absence of use of power" will result in companies' loss of power (Garriga and Melé, 2004, p. 56). Integrative social contract theory suggests an existence of a social contract linking business and society entailing indirect duties of business towards society (Donaldson, 1982). The concept of Corporate Citizenship theory involves the meaning of business citizens (Wood and Logsdon, 2001, 2002; Matten et al., 2003), signifying a purpose of belonging to a community (Garriga and Melé, 2004). The extent of corporate citizenship may differ. Limited citizenship has a similar nature to philanthropy, citizenship may also correspond to CSR, and an extended citizenship view is where companies step into the role of governments when they fail to protect citizens (Matten et al., 2003).

Integrative theories are the theories that discuss how corporations should fulfil and integrate social demands, given their reliance on society for their own success and prosperity (Garriga and Melé, 2004). This group of theories has subsets, including *issues management*, the *principle of public responsibility*, *stakeholder management*, and *corporate social performance*. Issues and issue management centre on how corporations respond to social and political issues that may significantly impact them (Wartick and Rude, 1986), and are thus considered to be of strategic importance (Greening and Gray, 1994). The principle of public responsibility defines the primary involvement of businesses in aspects such as economic tasks and legal requirements, and secondary involvement is a consequence of primary activities (Preston and Post, 1981). Furthermore, business should influence public policy processes when public policy has not been established or is being developed. Stakeholder management relates to actions carried out to influence those being affected or those that can affect business operation, such as through dialogue (Freeman, 1984). Corporate social performance (CSP) theory (Donaldson, 1982; Swanson, 1995) has been used as a way of gaining social legitimacy. It relates to the integrative social contract theory and integrates legitimacy theory, stakeholder theory, and institutional theory (Fernando and Lawrence,

2014). Of these, legitimacy theory has been defined as "a generalized perception or assumption that the actions of an entity are desirable, proper, or appropriate within some socially constructed system of norms, values, beliefs, and definitions" (Suchman, 1995, p. 574). Elements of corporate social performance theory can, for example, be seen as CSR principles, corporate social responsiveness processes, and outcomes of corporate behaviour (Wood, 1991).

Ethical theories are used to explain the ethical responsibilities of corporations in society and how they should assume responsibility based on ethical values. This is done through a normative stakeholder theory, and focuses on universal rights, sustainable development, a common-good approach (Garriga and Melé, 2004), and business ethics (Epstein, 1987; Carroll, 1999). The normative stakeholder theory highlights the situation where firms simultaneously take into consideration interests of all relevant stakeholders while balancing their multiple interests, not only the interests of the firm's shareholders (Garriga and Melé, 2004). Universal rights are for instance presented in the United Nations Global Compact, which includes ten principles in the areas of human rights, labour, the environment, anti-corruption, and human rights declarations (Garriga and Melé, 2004), and in the Arctic context more specifically declarations and conventions protecting the rights of Indigenous Peoples. Sustainable development is commonly defined as "development that meets the needs of the present without compromising the ability of future generations to meet their own needs" (World Commission on Environment and Development, 1987, p. 8). The role of businesses in supporting sustainable development is for instance presented in the Sustainable Development Goals, including Goal 8, decent work and economic growth, Goal 9, industry, innovation and infrastructure, and Goal 12, responsible consumption and production (United Nations, n.d.e). The common-good approach "holds the common good of society as the referential value for CSR" (Mahon and McGowan, 1991; Velasquez, 1992). Finally, the concept of business (or corporate) ethics is of relevance, but can be defined as "the study of business situations, activities, and decisions where issues of right and wrong are addressed" (Crane and Matten, 2007, p. 5).

Further resources

The concepts and theories discussed in this chapter are by no means all-inclusive of all the theoretical discussion about CSR since the middle of the last century, but still provide an important foundation for understanding the topic in an Arctic context. A few additional papers and publications will be mentioned, including CSR publications related to particular industries that are important in the Arctic context. These include *Corporate Social Responsibility as Institution: A Social Mechanisms Framework* (Bice, 2017) where the focus in on the global mining industry, and *Corporate Social Responsibility and the Oil Industry in the Russian Arctic: Global Norms and Neo-Paternalism* (Henry et al., 2016, p. 1340) focusing on the "gap between corporate social responsibility (CSR) as articulated in corporate offices and implemented at the local level". The local perception of CSR for Arctic petroleum in the Barents region has also been studied, suggesting "that those who gain directly from the petroleum industry and do not directly experience negative impacts were more inclined to be positive about the industry, although overall, general support for petroleum activity was high" (Kelman et al., 2016, p. 152). The acceptance of stakeholders, and businesses' rights to operate, has also been studied in relation to the social license to operate in mining operations in the Arctic, suggesting that transparency in mining operations is critical for an effective social license to operate (Heikkinen et al., 2016). The World Maritime University (WMU) publishes a book series on maritime and shipping affairs, including a book titled *Sustainable Shipping in a Changing Arctic* which addresses important Arctic-relevant subjects such as sustainable business development and ocean governance (Hildebrand et al., 2018), and another one focusing in general on *Corporate Social Responsibility in the Maritime Industry* (Froholdt, 2018), addressing environmental, legal, social, and economic aspects of the topic. The concept of energy justice has been discussed in relation to energy infrastructure development in the Arctic, drawing attention to "vulnerabilities of mis- and under-represented people" in the case of "fair and equitable infrastructural expansion" globally and locally (McCauley et al., 2016). Yet another marine-related publication worth mentioning is titled *The Shipping Industry, Ocean Governance and Environmental Law in the Paradigm Shift: In Search of a Pragmatic Balance for the Arctic* (Johansson and Donner, 2015). It claims that "CSR is intrinsically connected to ocean governance and 'stakeholder theory' is an authoritative concept in relation to the Arctic" (p. ix). The discussion is, however, limited to the marine environment and does not include human elements such as safety and security.

Case study: Albatros Expeditions

Albatros Expeditions (n.d.a), founded in 2017, is a Scandinavian company which provides year-round small ship expeditions in the Arctic and Antarctic. The expedition tours range from 6–14 days. Locations include the Arctic, Greenland, Svalbard, and Iceland, in some cases going into untouched places such as the Arctic Islands of the North Atlantic. The mother company, Albatros Travel Group, claims to place great emphasis on social responsibility in terms of strategy, implementation, and sustainability reports. Among the initiatives of the company include preventing plastic pollution from ships, recycling and reducing chemical usage, offering sustainable seafood, providing local support in places that are visited, having a "global positive impact to our oceans, seabirds and marine species, as well as conservation of our planet", and generating cash flow in local communities such as through collaboration with small local businesses (Albatros Expeditions, 2019). The company is a member of the Arctic Expedition Cruise Operators (AECO) and the International Association of Antarctica Tour Operators (IAATO). The aim of the company is to offer additional sustainability training to the onboard lectures provided by experts to expedition guests and members of the expedition team, and information packages including information about conservation and safety. These lectures focus on the natural environment, cultural differences in the region, and donations to specific projects (Albatros Expeditions, 2019). The Albatros Expeditions CSR is published online, categorized into people, planet, and profit. The first theme centres around human rights and adverse human rights impacts, the second theme is on sustainability, and the third one is on actions against corruption, stakeholder engagement, and the precautionary apporach to the natural environment (Albatros Expeditions, n.d.b).

Questions for discussion

1. Which CSR concepts, frameworks, and theories are of relevance for the Albatros Expeditions?
2. In order to protect the intangible social license to operate in the Arctic, are there some critical elements of CSR missing? If so, which ones?
3. Why is the concept of risk of importance when discussing CSR concepts, frameworks, and theories in an Arctic context?
4. How, if at all, is the concept of risk articulated by the Albatros Expeditions?

Chapter summary

Historical milestones, concepts, frameworks, and theories of relevance for the modern CSR dialogue, sustainability, and Indigenous Peoples have been discussed in this chapter. A timeline of the historical milestones in CSR was presented, showing the initial modern phase of CSR (1953–1969), the early modern CSR phase and institutional debate (1970–1979), the period of early operationalization of CSR (1980–1989), the period of later operationalization of CSR (1990–1990), the period of mature, strategic, and institutionalized approaches to CSR (2000–2009), and the period of reporting, disclosure, and results (2010–2019). CSR concepts were then discussed, both those developed by academic scholars and those originating from practitioners and policymakers. Other concepts interrelated to, and overlapping with the CSR concept, were also mentioned, these being accountability, business ethics, corporate citizenship, environmental sustainability, integrated or CSR reporting, stakeholder theory, sustainable development and business/corporate sustainability, social license to operate, and shared value. Subsequently, three conceptual CSR frameworks, and an Arctic-specific CSR framework were put forward: Carroll's pyramid of CSR, the intersecting circles of CSR, concentric circles of CSR, and the Arctic CSR and sustainability framework. CSR theories discussed in the chapter are categorized as instrumental theories, political theories, integrative theories, and ethical theories. The case studied in the chapter is Albatros Expeditions, a Scandinavian company which provides year-round small ship expeditions in the Arctic and Antarctic. This case offers the opportunity of applying the concepts, frameworks, and theories introduced in the chapter to a real-life case.

References

AccountAbility. (2008). Standards. Accessed on 10 March 2020 from www.accountability.org/standards/

Albatros Expeditions. (2019). *Communication of Progress 2018–2019*. Accessed on 19 March 2020 from https://albatros-expeditions.com/sites/default/files/files/COP%20AEX%202019.pdf

Albatros Expeditions. (n.d.a). Arctic. Accessed on 19 March 2020 from https://albatros-expeditions.com/expeditions/arctic

Albatros Expeditions. (n.d.b). CSR. Accessed on 19 March 2020 from https://albatros-expeditions.com/csr

Arctic Council. (1996). Declaration on the Establishment of the Arctic Council. Ottawa: Representatives of the Arctic States.

Arctic Council. (2013a). Kiruna Declaration. Accessed on 14 February 2020 from https://oaarchive.arctic-council.org/handle/11374/93

Arctic Council. (2013b). CSR in the Arctic – way forward. Accessed on 16 March 2020 from https://oaarchive.arctic-council.org/bitstream/handle/11374/1221/Doc5-1_CSR_Website_layout.pdf?sequence=1&isAllowed=y

Arctic Council. (n.d.). Corporate social responsibility in the Arctic. Accessed on 18 March 2020 from www.sdwg.org/corporate-social-responsibility-in-the-arctic/

Arctic Economic Council. (2014). About us. Accessed on 16 March 2020 from https://arcticeconomiccouncil.com/about-us/

Bice, S. (2017). Corporate social responsibility as institution: A social mechanisms framework. *Journal of Business Ethics*, 142, 17–34. doi:10.1007/s10551-015-2791-1

Bowen, H.R. (1953). *Social Responsibilities of the Businessman*. Iowa City, IA: University of Iowa Press.

Business for Social Responsibility (BSR). (n.d.). Business for Social Responsibility (BSR). Accessed on 5 March 2020 from www.devex.com/organizations/business-for-social-responsibility-bsr-49469

Burke, L. and Logston, J.M. (1996). How corporate responsibility pays off. *Long Range Planning*, 29(4), 495–502. https://doi.org/10.1016/0024-6301(96)00041-6

Carlson, R. (1962). *Silent Spring*. Boston, MA: Houghton Mifflin Company.

Carroll, A.B. (1979). A three-dimensional conceptual model of corporate performance. *The Academy of Management Review*, 4(4), 497–505. doi:10.2307/257850

Carroll, A.B. (1991). The pyramid of corporate social responsibility: Toward the moral management of organizational stakeholders. *Business Horizons*, 34(4), 39–48.

Carroll, A.B. (1999). Corporate social responsibility. *Business & Society*, 38(3), 268–295.

Carroll, A.B. (2008). A history of corporate social responsibility: Concepts and practices. In A.M. Andrew Crane, D. Matten, J. Moon, and D. Siegel (Eds.), *The Oxford Handbook of Corporate Social Responsibility* (pp. 19–46). New York: Oxford University Press.

Carroll, A.B. (2016). Corporate social responsibility: The centrepiece of competing and complementary frameworks. *Organizational Dynamics*, 44(2), 87–96.

CDP (2020). About us. What we do. Accessed on 10 March 2020 from www.cdp.net/en/info/about-us/what-we-do

Chandler, D. (2016). *Strategic Corporate Social Responsibility: Sustainable Value Creation*. Thousand Oaks, CA: SAGE Publications.

Chandler, D. and Werther, W.B. (2013). *Strategic Corporate Social Responsibility: Stakeholders, Globalization, and Sustainable Value Creation* (3rd ed.). Thousand Oaks: SAGE Publications.

Cochran, P.L. and Wood, R.A. (1984). Corporate social responsibility and financial performance. *Academy of Management Journal*, 27(1), 42–56. https://doi.org/10.2307/255956

Committee for Economic Development. (1971). *Social Responsibilities of Business Corporations*. New York: Author.

Corporate Sustainability Assessment (n.d.). DJSI index family. Accessed on 16 March 2020 from www.robecosam.com/csa/indices/djsi-index-family.html

Crane, A. and Matten, D. (2007). *Business Ethics: Managing Corporate Citizenship and Sustainability in the Age of Globalization* (2nd ed.). Oxford, UK: Oxford University Press.

Davis, K. (1960, Spring). Can business afford to ignore social responsibilities? *California Management Review*, 2(3), 70–76.

Davis, K. (1967). Understanding the social responsibility puzzle. *Business Horizons*, 10(4), 45–51. https://doi.org/10.1016/0007-6813(67)90007-9

Davis, K. and Blomstrom, R.L. (1966). *Business and Its Environment*. New York: McGraw Hill.

Donaldson, T. (1982). *Corporations and Morality*. Englewood Cliffs, NJ: Prentice Hall.

Earth Day. (n.d.). The history of the Earth Day. Accessed on 5 March 2020 from www.earthday.org/history/

Epstein, E.M. (1987). The corporate social policy process: Beyond business ethics, corporate social responsibility, and corporate social responsiveness. *California Management Review*, 29, 99–114.

European Commission. (2002). *Communication from the Commission concerning Corporate Social Responsibility: A Business Contribution to Sustainable Development* (COM(2002) 347 final). Brussels: Author.

European Commission. (2003). *Council Resolution of 6 February 2003 on Corporate Social Responsibility* (2003/C 39/02). Brussels: Author. Accessed on 27 December 2020 from https://eur-lex.europa.eu/LexUriServ/LexUriServ.do?uri=OJ:C:2003:039:0003:0004:EN:

European Commission. (2006). *Communication from the Commission to the European Parliament, the Council and the European Economic and Social Committee Implementing the Partnership for Growth and Jobs: Making Europe a Pole of Excellence on Corporate Social Responsibility*. Brussels: Author. Accessed on 2 April 2021 from https://eur-lex.europa.eu/LexUriServ/LexUriServ.do?uri=COM:2006:0136:FIN:en:PDF

European Commission. (2011). New EU definition on CSR mirrors Enterprise 2020 aspirations. Accessed on 17 March 2020 from www.csreurope.org/new-eu-definition-csr-mirrors-enterprise-2020-aspirations

European Commission. (2019). *Commission Staff Working Document: Corporate Social Responsibility, Responsible Business Conduct, and Business & Human Rights: Overview of Progress* (SWD(2019) 143 final). Brussels: Author.

European Commission. (n.d.). Corporate social responsibility & responsible business conduct. Accessed on 17 March 2020 from https://ec.europa.eu/growth/industry/sustainability/corporate-social-responsibility_en

Frederick, W.C. (1960). The growing concern over business responsibility. *California Management Review*, 2, 54–61.

Fernando, S. and Lawrence, S. (2014). A theoretical framework for CSR practices: Integrating legitimacy theory, stakeholder theory and institutional theory. *The Journal of Theoretical Accounting*, 10(1), 149–178.

Freeman, E.R. (1984). *Strategic Management: A Stakeholder Approach*. Boston, MA: Pitman.

Friedman, M. (1970, September). The social responsibility of business is to increase its profits. *New York Times Magazine*, p. 32. Accessed on 2 April 2021 from www.nytimes.com/1970/09/13/archives/a-friedman-doctrine-the-social-responsibility-of-business-is-to.html

Froholdt, L.L. (Ed.) (2018). *Corporate Social Responsibility in the Maritime Industry*. (WMU Studies in Maritime Affairs Book 5). New York: Springer International Publishing AG.

Garriga, E. and Melé, D. (2004). Corporate social responsibility theories: Mapping the territory. *Journal of Business Ethics*, 53, 51–71. https://doi.org/10.1023/B:BUSI.0000039399.90587.34

Geva, A. (2008). Three models of corporate social responsibility: Interrelationships between theory, research, and practice. *Business and Society Review*, 113(1), 1–41. doi:10.1111/j.1467-8594.2008.00311.x

Global Reporting Initiative (GRI). (n.d.). GRI's history. Accessed on 10 March 2020 from www.globalreporting.org/information/about-gri/gri-history/Pages/GRI's%20history.aspx

Greening, D.W. and Gray, B. (1994). Testing a model of organizational response to social and political issues. *Academy of Management Journal*, 37, 467–498. www.jstor.org/stable/256697

Heikkinen, H.I., Lépy, É., Sarkki, S., and Komu, T. (2016). Challenges in acquiring a social licence to mine in the globalising Arctic. *Polar Record*, 52(4), 399–411. https://doi.org/10.1017/S0032247413000843

Henry, L.A., Nysten-Haarala, S., Tulaeva, S., and Tysiachniouk, M. (2016). Corporate social responsibility and the oil industry in the Russian Arctic: Global norms and neo-paternalism. *Europe-Asia Studies*, 68(8), 1340–1368. http://dx.doi.org/10.1080/09668136.2016.1233523

Heslin, P.A. and Ochoa, J.D. (2008). Understanding and developing strategic corporate social responsibility. *Organizational Dynamics*, 37(2), 125–144 https://ec.europa.eu/growth/industry/sustainability/corporate-social-responsibility_en

Hildebrand, L.P., Bringham, L.W., and Johansson, T.M. (Eds.) (2018). *Sustainable Shipping in a Changing Arctic* (WMU Studies in Maritime Affairs Book 7). New York: Springer International Publishing AG.

Husted, B.W. and Allen, D.B. (2007). Strategic corporate social responsibility and value creation among large firms: Lessons from the Spanish experience. *Long Range Planning*, 40(6), 594–610.

International Integrated Reporting Council. (2010). Formation of the International Integrated Reporting Committee (IIRC). Accessed on 16 March 2020 from https://integratedreporting.org/news/formation-of-the-international-integrated-reporting-committee-iirc/

International Labour Organization (ILO). (1957). Convention No. 107. Indigenous and Tribal Peoples. Accessed on 5 March 2020 from www.ilo.org/dyn/normlex/en/f?p=NORMLEXPUB:12100:0::NO::P12100_ILO_CODE:C107

International Labour Organization (ILO). (1989). Convention No. 169. Indigenous and Tribal Peoples Convention. Accessed on 16 March 2020 from www.ilo.org/dyn/normlex/en/f?p=NORMLEXPUB:12100:0::NO::P12100_ILO_CODE:C169

International Nuclear Safety Advisory Group. (1992). *The Chernobyl Accident: Updating of INSAG-1*. (Safety Series No. 75-INSAG-7. Vienna: Author.

ISO. (1996). ISO 14001:1996. Environmental management systems. Accessed on 2 April 2021 from www.iso.org/standard/23142.html

ISO. (n.d.). ISO 26000:2010 (en). Accessed on 16 March 2020 from www.iso.org/obp/ui/#iso:std:iso:26000:ed-1:v1:en

Jensen, M.C. (2000). Value maximization, stakeholder theory, and the corporate objective function. *Business Ethics Quarterly*, 12(2), 235–256. www.jstor.org/stable/3857812

Jensen, M.C. and Meckling, W. (1976). Theory of the firm: Managerial behavior, agency cost, and capital structure. *Journal of Financial Economics*, 3(October), 305–360. https://doi.org/10.1016/0304-405X(76)90026-X

Johannsdottir, L. and Cook, D. (2019). Systemic risk of maritime-related oil spills viewed from an Arctic and insurance perspective. *Ocean & Coastal Management*, 179(1): 104853. doi:10.1016/j.ocecoaman.2019.104853

Johansson, T. and Donner, P. (2015). *The Shipping Industry, Ocean Governance and Environmental Law in the Paradigm Shift: In Search of a Pragmatic Balance for the Arctic*. Heidelberg, New York, Dordrecht, London: Springer Cham.

Johnson, H.L. (1971). *Business in Contemporary Society: Framework and Issues*. Belmont, CA: Wadsworth.

Jones, T.M. (1980). Corporate social responsibility revisited, redefined. *California Management Review*, 22(3), 59–67.

Kelman, I., Loe, J.S.P., Rowe, E.W., Wilson, E., Poussenkova, N., Nikitina, E., and Fjærtoft, D.B. (2016). Local perceptions of corporate social responsibility for Arctic petroleum in the Barents Region. *Arctic Review on Law and Politics*, 7(2), 152–178. https://doi.org/10.17585/arctic.v7.418

Khandker, S., Khalily, B., and Khan, Z. (1994).. *Is Grameen Bank Sustainable?* Washington, D.C.: World Bank, Human Resources Development and Operations Policy.

Lantos, G.P. (2001). The boundaries of strategic corporate social responsibility. *Journal of Consumer Marketing*, 18(7), 595–632.

Latapí Agudelo, M.A., Johannsdottir, L., and Davidsdottir, B. (2019). Literature review of the history and evolution of corporate social responsibility. *International Journal of Corporate Social Responsibility*, 4(1), 1–23. https://doi.org/10.1186/s40991-018-0039-y

Mahon, J.F. and McGowan, R.A. (1991). Searching for the common good: A process-oriented approach. *Business Horizons*, 34(4), 79–87. https://doi.org/10.1016/0007-6813(91)90010-S

Margolis, J.D. and Walsh, J.P. (2003). Misery loves companies: Rethinking social initiatives by business. *Administrative Science Quarterly*, 48(2): 268–305. doi:10.2307/3556659

Marrewijk, M. (2003). Concepts and definitions of CSR and corporate sustainability: Between agency and communion. *Journal of Business Ethics*, 44, 95–105. https://doi.org/10.1023/A:1023331212247

Matten, D., Crane, A., and Chapple, W. (2003). Behind the mask: Revealing the true face of corporate citizenship. *Journal of Business Ethics*, 45(1–2), 109–120.

McCauley, D., Heffron, R., Pavlenko, M., Rehner, R., and Holmes, R. (2016). Energy justice in the Arctic: Implications for energy infrastructural development in the Arctic. *Energy Research & Social Science*, 16, 141–146. http://dx.doi.org/10.1016/j.erss.2016.03.019

McGuire, J. (1963). *Business and Society*. New York: McGraw Hill.

Meadows, D.H., Meadows, D.L., Randers, J., and Behrens III, W.W. (1972). *The Limits to Growth: A Report for the Club of Rome's Project on the Predicament of Mankind*. New York: Universe Books.

Nasdaq. (2019). *ESG Reporting Guide 2.0: A Support Resource for Companies*. Accessed on 16 March 2020 from www.nasdaq.com/docs/2019/11/26/2019-ESG-Reporting-Guide.pdf

Office of the High Commissioner for Human Rights. (n.d.). *Mandate of the Working Group on Indigenous Populations*. Geneva: Author.

Organisation for Economic Co-operation and Development (OECD). (2011). The OECD guidelines. Accessed on 16 March 2020 from www.oecdwatch.org/oecd-ncps/the-oecd-guidelines-for-mnes/

Organisation for Economic Co-operation and Development (OECD). (n.d.). *Celebrating 40 Years of the OECD Environment Policy Committee 1971–2011*. Paris: Author. Accessed on 20 December 2020 from www.oecd.org/env/48943696.pdf

Orlitzky, M., Schmidt, F.L., and Rynes, S.L. (2003). Corporate social and financial performance: A meta-analysis. *Organization Studies*, 24(3), 403–441. https://doi.org/10.1177/0170840603024003910

Porter, M.E. and Kramer, M.R. (2006). Strategy & society. *Harvard Business Review*, December, 1–16.

Porter, M.E. and Kramer, M.R. (2011). Creating shared value. *Harvard Business Review* (January–February), 1–17.

Preston, L.E. and Post, J.E. (1981). Private management and public policy. *California Management Review*, 23(3), 56–63. https://doi.org/10.2307/41172602

Principles for Responsible Investment (PRI). (n.d.). About the PRI. Accessed on 10 March 2020 from www.unpri.org/pri/about-the-pri

Ross, S. (1973). The economy theory of the agency: The principal's problem. *American Economic Review*, 63, 134–139. www.jstor.org/stable/1817064

Schmidheiny, S. and BCSD. (1992). *Changing Course: A Global Business Perspective on Development and the Environment*. Cambridge, MA: MIT Press.

Schwartz, M.S. and Carroll, A.B. (2003). Corporate social responsibility: A three-domain approach. *Business Ethics Quarterly*, 13(4), 503–530. https://doi.org/10.5840/beq200313435

Sethi, S.P. (1975, Spring). Dimensions of corporate social performance: An analytic framework. *California Management Review*, 17, 58–64.

Skinner, S.K. and Reilly, W.K. (1989). *The Exxon Valdez Oil Spill: A Report to the President*. Washington, D.C.: National Response Team. Accessed on 2 April 2021 from https://archive.epa.gov/epa/aboutepa/exxon-valdez-oil-spill-report-president-executive-summary.html

Steiner, G.A. (1971). *Business and Society* (1st ed). New York: Random House. Accessed on 20 December 2020 from https://trove.nla.gov.au/work/21398855

Stockholm Resilience Centre. (2009). The nine planetary boundaries. Accessed on 10 March 2020 from www.stockholmresilience.org/research/researchnews/tippingtowardstheunknown/thenineplanetaryboundaries.4.1fe8f33123572b59ab80007039.html

Suchman, M.C. (1995). Managing legitimacy: Strategic and institutional approaches. *Academy of Management Journal*, 20(3), 571–610. www.jstor.org/stable/258788

Sustainability Accounting Standards Board. (n.d.). Mission. Accessed on 16 March 2020 from www.sasb.org/governance/

Swanson, D. L. (1995). Addressing a theoretical problem by reorienting the corporate social performance model. *The Academy of Management Review*, 20(1), 43–64. doi:10.2307/258886

The International Institute for Sustainable Development (2012). *Sustainable Development Timeline*. Accessed on 5 March 2020 from www.iisd.org/pdf/2012/sd_timeline_2012.pdf

The Nobel Prize. (2007). The Nobel Peace Prize 2007. Accessed on 10 March 2020 from www.nobelprize.org/prizes/peace/2007/summary/

Trapp, N.L. (2012). Corporation as climate ambassador: Transcending business sector boundaries in a Swedish CSR campaign. *Public Relations Review*, 38(3), 458–465.

Union of International Association. (n.d.). United Nations Commission on Sustainable Development (CSD). Accessed on 16 March 2020 from https://uia.org/s/or/en/1100043826

United Nations. (2011). *Guiding Principles on Business and Human Rights* (HR/PUB/11/04). New York and Geneva: Human Rights Office of the High Commissioner.

United Nations. (n.d.a). Historical overview. Accessed on 5 March 2020 from www.un.org/development/desa/indigenouspeoples/declaration-on-the-rights-of-indigenous-peoples/historical-overview.html

United Nations. (n.d.b). Background. Accessed on 10 March 2020 from www.un.org/millenniumgoals/bkgd.shtml

United Nations. (n.d.c). Conferences and reports on the environment. Accessed on 16 March 2020 from https://research.un.org/en/docs/environment/conferences

United Nations. (n.d.d). History of the Kyoto Protocol. Accessed on 16 March 2020 from https://unfccc.int/process-and-meetings/the-kyoto-protocol/history-of-the-kyoto-protocol

United Nations. (n.d.e). Sustainable Development Goals. Accessed on 16 March 2020 from https://sustainabledevelopment.un.org/?menu=1300

United Nations Environment Programme. (n.d.). About Montreal Protocol. Accessed on 16 March 2020 from www.unenvironment.org/ozonaction/who-we-are/about-montreal-protocol

United Nations Framework Convention on Climate Change (UNFCCC). (n.d.). Timeline. Accessed on 10 March 2020 from https://unfccc.int/timeline/

United Nations Global Compact. (n.d.). About the UN Global Compact. Accessed on 10 March 2020 from www.unglobalcompact.org/about

Vammen Larsen, S., Bors, E., Johannsdottir, L., Gladun, E., Gritsenko, D., Nysten-Haarala, S., Tulaeva, S., and Sformo, T. (2019). A conceptual framework of Arctic economies for policy-making, research, and practice. *Global Policy*, 10(4), 686–696. doi:10.1111/1758-5899.12720

Velasquez, M. (1992). International business, morality and the common good. *Business Ethics Quarterly*, 2(1), 27–40. www.jstor.org/stable/3857220

Walton, C.C. (1967). *Corporate Social Responsibilities*. Belmont, CA: Wadsworth.

Wartick, S.L. and Rude, R.E. (1986). Issues management: Corporate fad or corporate function? *California Management Review*, 29(1), 124–132. https://doi.org/10.2307/41165231

Werther, W.B. and Chandler, D. (2005). Strategic corporate social responsibility as global brand insurance. *Business Horizons*, 48(4), 317–324.

Wilson, M. (2003). Corporate sustainability: What is it and where does it come from? *Ivey Business Journal*, March/April, 1–5.

Wood, D.J. and Lodgson, J.M. (2002). Business citizenship: From individuals to organizations. *Business Ethics Quarterly, Ruffin Series*, No. 3, 59–94. doi:10.5840/ruffinx200232

Wood, D.J. (1991). Corporate social performance revisited. *The Academy of Management Review*, 16(4), 691–718.

Wood, D.J. and Logsdon, J.M. (2001). Theorizing business citizenship. In J. Andriof and M. McIntosh (Eds.), *Perspectives on Corporate Citizenship* (pp. 83–103). Sheffield: Greenleaf.

World Business Council for Sustainable Development (WBCSD). (n.d.). Our history. Accessed on 16 March 2020 from www.wbcsd.org/Overview/Our-history

World Commission on Environment and Development. (1987). *Our Common Future*. Oxford, UK: Oxford University Press.

World Economic Forum. (2015). *Arctic Investment Protocol: Guidelines for Responsible Investment in the Arctic* (REF 161115). Geneva: Author.

World Trade Organization. (1995). Overview of developments in international trade and the trading system. Accessed on 16 March 2020 from www.wto.org/english/news_e/pres95_e/ov11.htm

3 Relevance of CSR in a world in transition

Contemporary trends

Learning objectives

- Recognize climate change impacts in the Arctic.
- Understand the underlying climate-related concepts.
- Recognize the pros and the cons of Arctic energy development, infrastructure, transportation, and energy utilization.
- Identify international agreements that draw attention to characteristics of resilience.
- Understand how climate change is influencing contemporary CSR trends.

Introduction

For several decades there has been transformative change in Arctic climatic conditions (IPCC, 2014). Environmental and ecological transformations in the Arctic are largely driven by economic activities, by and large taking place outside the Arctic (Arctic Council, 2016). The Arctic climate is changing at a faster rate than that in mid-latitudes (Overland et al., 2013), and the magnitude of Arctic-wide warming, documented over 30 years, shows a warming trend of 1.9 °C, a rate that is three times higher than the global average (Comiso and Hall, 2014). Changes are observed in the Arctic sea-ice and coral-reef systems, and currently Arctic ecosystems are undergoing irreversible regime shifts. The risk of losing biodiversity, marine and coastal ecosystems, and ecosystem services is also a risk for coastal livelihoods, such as in case of fishing communities. This therefore risks the health and well-being of residents in the Arctic as change in the physical environment causes food insecurity, lack of safe and reliable drinking water, and may cause damage to infrastructure, for instance in permafrost regions. Impacts on livelihoods of Indigenous Peoples are expected to go beyond the effects of economic and socio-political changes (IPCC, 2014). These changes call for both mitigation and adaptation policies and actions to ensure the resilience of human and non-human Arctic systems.

Business opportunities in the Arctic, resulting from climate change, are also recognized. Previously inaccessible resources are becoming accessible, for instance mineral resources, developed through mining and oil and gas projects. Retreating of sea ice brings about shorter shipping routes, which are of interest for transportation of global goods as well as in the case of large-scale projects carried out within the Arctic region. This results in increasing shipping and logistics within and through Arctic waters, but less sea ice is also influencing Arctic tourism. Impacts on fishing are mixed, as fishing might in some cases boom but long-term climate impacts of particular fish stocks may also be negative. In order to realize these business opportunities, businesses involved need to manage unique and significant Arctic risks (Emmerson and Lahn, 2012), and take into account and manoeuvre different scales of risks (Johannsdottir and Cook, 2019). Other questions concerning Arctic business opportunities are of a political, geopolitical, and legal nature, such as: *Who owns what? Who controls what?* Since these projects entail risks to the natural environment, there are also questions of importance in the case of disasters, such as oil spills: *Who are liable? Under what liability regime are they liable?* (Emmerson and Lahn, 2012).

Climate change, as well as utilization of new economic opportunities, are having and will continue to have significant socio-economic impacts within the Arctic region. In this chapter, climate change, energy development, adaptation, and resilience to change are discussed, as well as the contemporary trends on strategic corporate social responsibility (CSR) in a world in transition.

Types of discussion and issues explored in Chapter 3

This section provides a short overview of climate-related concepts that are of relevance for the chapter's discussion. The definitions used are mainly those adopted by the Intergovernmental Panel on Climate Change (IPCC) and the Arctic Council. Climate change impacts in the Arctic are then briefly discussed, moving into discussing energy development, infrastructure, transportation, and energy utilization. This discussion also focuses on scales of risks that are of importance when Arctic industrial projects are considered and carried out. Climate change adaptation is then discussed in the context of ecological and socio-economic changes. The resilience concept is then examined, such as in the context of the Arctic Council's reports, the Paris Agreement, the Sustainable Development Goals, and the Sendai Framework for Disaster Risk Reduction: all of these major international agreements from 2015 draw attention to characteristics of resilience (Arctic Council, 2016), including CSR and the role of businesses in the Arctic. The idea of moving from building systems resilience to thriving communities is also investigated. The case investigated in this chapter is the role of financial institutions and why some of them no longer support investments in non-renewable energy projects.

Background: Climate-related concepts

Some of the main climate-related concepts used in this chapter are mitigation, adaptation, resilience, socio-economic system, tipping point (irreversibility), and socio-ecological system, and as such need to be clarified. Others will be introduced where relevant to the discussion. The first concept, mitigation of climate change, has been defined by the Intergovernmental Panel on Climate Change (IPCC), an intergovernmental body of the United Nations. Mitigation requires "[a] human intervention to reduce emissions or enhance the sinks of greenhouse gases" (IPCC, n.d.). The second concept, adaptation, has been defined both for human and natural systems. The former refers to "the process of adjustment to actual or expected climate and its effects, in order to moderate harm or exploit beneficial opportunities". The latter is defined as "the process of adjustment to actual climate and its effects; human intervention may facilitate adjustment to expected climate and its effects" (IPCC, n.d.). This relates to concepts such as adaptability, adaptive capacity, adaptation options, and adaptation pathways. The third concept, resilience, is based on a definition used by the Arctic Council (2013a). It refers to the "capacity of social, economic and environmental systems to cope with a hazardous event or trend or disturbance, responding or reorganizing in ways that maintain their essential function, identity and structure while also maintaining the capacity for adaptation, learning and transformation" (IPCC, n.d.). The fourth concept, socio-economic system, originates from the Arctic Council (2016) and Berkes and Folke (1998) and is referred to as an "integrated system that includes human societies and ecosystems, in which humans are part of nature. The functions of such a system arise from the interactions and interdependence of the social and ecological subsystems. The system's structure is characterized by reciprocal feedback, emphasising that humans must be seen as a part of, not apart from, nature" (IPCC, n.d.). The fifth concept, tipping point, suggests "[a] level of change in system properties beyond which a system reorganizes, often abruptly, and does not return to the initial state even if the drivers of the change are abated. For the climate system, it refers to a critical threshold when global or regional climate changes from one stable state to another stable state" (IPCC, n.d.). Finally, socio-ecological systems include both "human and natural systems in any given place", for instance ecosystems supporting local communities (Arctic Council, 2016, p. ix).

Climate change in the Arctic

Carbon emissions resulting from human activities are related to fossil fuel energy production and consumption around the world, but global emissions accelerate the changes happening in the Arctic, causing impacts on socio-economic systems and socio-ecosystems; these impacts are not just felt in the Arctic but around the globe (Arctic Council, 2016; Arruda and Krutkowski, 2017). These changes need to be decelerated through the implementation of mitigation actions, and resilience capacity built up, so people and ecosystems can adapt (Arctic Council, 2016) and hopefully thrive. The IPCC's 2007 (Anisimov et al., 2007) assessment of the Polar Regions claims with very high confidence that there will be ongoing impacts on freshwater and terrestrial species, communities, and ecosystems, and with high confidence that such changes will continue with consequences for biological resources. There is plentiful evidence on these impacts: on a range of species and tree lines; methane emissions; warming and thawing of permafrost; flows of freshwater and ice into polar oceans impacting the sea level; and a decline in glaciers including the Greenland ice sheet; acidification of polar waters with adverse consequences on calcified organisms, and subsequent negative impacts on species relying upon them (Anisimov et al., 2007). The frequency of extreme weather may increasingly cause events such as forest fires or flooding (Mack et al., 2011). These changes are seen as a global early-warning signals, since the warming in the Arctic is happening more rapidly than in other parts of the world (Overland et al., 2013; Comiso and Hall, 2014).

Impacts will be region-specific and will vary widely. For instance, impacts on fisheries might be beneficial in some cases, but adverse in other cases. Adverse impacts will be on ice-dependent marine wildlife, but accessibility and availability of food may also be affected. In the case of hunting and travelling, personal safety might be at risk, thus having an impact on people's livelihoods; the resilience of Indigenous Peoples may be put to the test. Community infrastructure will be at risk due to thawing of the permafrost, calling for investments to relocate communities and tangible structures (Anisimov et al., 2007), such as buildings, bridges, roads, railways, and pipelines; new types of design and adaptation will be required (Emmerson and Lahn, 2012). Anecdotal evidence includes the disappearing of Indigenous People's livelihoods, and uninhabitable local villages (Arctic Council, 2016), and many inland areas will be less accessible (Emmerson and Lahn, 2012).

These changes are not just negative as economic benefits are also predicted. These include the potential for enhanced hydropower and oil and gas production, resulting in reduction in heating costs; greater marine access and access to mineral resources; opening of new northern sea routes; benefits for shipping, logistics and Arctic tourism; and opportunities in agriculture and forestry. Considerable investment is needed if economic opportunities are to be utilized (Anisimov et al., 2007; Emmerson and Lahn, 2012; Arctic Council, 2016).

Both the challenges and opportunities arising in the Arctic cause uncertainties in the local and geopolitical environment and Arctic governance. Questions such as how to decide who owns what and who controls what will arise, as well as who will be liable and under which liability regime in worst-case scenarios such as oil spills. There is therefore a need to manage Arctic risks through risk governance, risk mitigation, and risk transfer (Emmerson and Lahn, 2012; Arruda, 2014; Arruda, 2015; Johannsdottir and Cook, 2019). It should also be noted that Arctic conditions and solutions need to be viewed in a holistic and systemic manner as Arctic systems are closely interconnected (Arctic Council, 2016; Johannsdottir and Cook, 2019), and the same applies to the changes in Arctic systems that "cascade across geophysical, ecological and human elements of social-ecological systems" (Arctic Council, 2016, p. x).

Arctic energy development, infrastructure, transportation, and energy utilization

The Carbon Majors Database of the Carbon Disclosure Project (CDP) incorporates 100 fossil fuel producers which are accountable for one trillion tonnes of greenhouse gas emissions, linked to more than 70% of the world's global industrial greenhouse gas emissions since 1988 (Griffin, 2017). It is of importance to discuss both non-renewable and renewable energy development in the Arctic. The former relates to extractive resources, including oil and gas, coal, natural gas, and methane accounting for extraction of around 20% of the world's currently utilized energy resources, and the latter comprises solar, wind, geothermal, hydro, and biomass, which are to date greatly unexplored for the purpose of meeting future energy demand (Arruda, 2019).

There has been growing interest in the Arctic mineral resources, in this context oil and gas, dependent on global market conditions of supply and demand, and investors' interest in such megaprojects (Emmerson and Lahn, 2012). Uncertainty also relates to infrastructure gaps in the region, as well as trade-offs between industries such as fisheries, reindeer herding, and oil and gas. Investment decisions therefore rely on political and regulatory conditions at the local, national, and global level, which are subject to change. Megaprojects of this nature are also exposed to various other risk factors, including geological, environmental, and operational risks (Emmerson and Lahn, 2012).

From a business perspective, decisions are based on feasibility, commercial attractiveness, and access. The potential awards of drilling in the Arctic are significant from the financial point of view, but push companies into "more extreme and fragile environments with ecosystems already under stress from the impacts of climate change" (Rees and Sharp, 2011, p. 20). Technologies used for exploration, drilling, and offshore production have amplified the prospects of finding oil and gas, and fewer installations are needed for extracting the resources (Emmerson and Lahn, 2012). The amount of oil and gas reserves in the Arctic and the future of energy supply is furthermore subject to environmental concerns. An assessment made by the United States Geological Survey suggests that undiscovered gas comprises around 30% of world's reserves, mostly concentrated in Russia, and oil comprises around 13%, mostly in offshore Norway and in the American Arctic. These reserves are mainly offshore where the depth of water is 500 metres or less (Gautier et al., 2009; Emmerson and Lahn, 2012).

Potential risks of non-renewable energy projects relate to drilling and shipping of hydrocarbon resources. These economic activities are risky given the harsh environment and working conditions in the Arctic, including geographic remoteness and isolation, lack of infrastructure, search and rescue and clean-up capabilities, extreme weather and/or ice conditions, costs of insurance coverage and unfavourable insurance terms and conditions, costs of icebreaker escort and self-insurance cost, and geopolitical tensions, just to name a few (Rees and Sharp, 2011; Emmerson and Lahn, 2012; Johannsdottir and Cook, 2015). Lloyd's of London has specifically discussed the challenges of drilling in extreme environments, claiming that as these activities move "further offshore and into remoter areas the operational, environmental and regulatory risks for oil and gas companies will grow significantly, with considerable implications for insurers". In addition, interference with fragile Arctic ecosystems "may have a negative impact on these environments and attract negative media attention, which can severely damage a company's reputation" (Rees and Sharp, 2011, pp. 6 & 8). Increased public awareness is recognized as a challenge, and that a rising number of claims for compensation exceed recovery and clean-up costs, as the impact of ecosystem losses on local economies also need to be compensated for (Rees and Sharp, 2011). Addressing Arctic-related business risks requires solutions such as having baseline knowledge, well-developed and functioning risk management frameworks and procedures, sharing of knowledge, cooperation, training, and use of technology most appropriate to Arctic conditions (Emmerson and Lahn, 2012; Johannsdottir and Cook, 2015).

Systemic risks of marine-related oil spills from drilling and shipping have been studied (Arruda, 2015; Johannsdottir and Cook, 2019). Findings include a hierarchy framework, see Figure 3.1. This shows different levels of risks: risk to subsistence communities; risk to enterprises, in this case oil and gas or shipping companies; portfolio risk, meaning groups of industries; systemic risk where breakdown of systems occur, such as in the case of oil spills; and existential risk, as in climate change, destruction of livelihoods, or currently COVID-19. When the risk of oil spills is viewed from the systemic level, there is an existential risk for those relying on subsistence resources for their livelihood and well-being, given the potential collapse of local ecosystems that Indigenous Peoples rely on (Johannsdottir & Cook, 2019). Therefore, a trade-off between different economic activities has been identified (Emmerson and Lahn, 2012; Johannsdottir and Cook, 2015), such as between fishing, tourism, and offshore oil and gas projects.

"Renewable energy can deliver half of all emission reductions needed to meet global targets" (Poelzer et al., 2016, p. 1); the role of the Arctic and sub-Arctic regions has been recognized in relation to renewable energy policies. This is evident, for instance, in the *U.S.-Canada Joint Statement on Climate, Energy, and Arctic Leadership* in support of the Paris Agreement (Prime Minister of Canada & The White House, 2016). However, the production and use of renewable energy differs greatly within the region. For instance, renewable energy is a much more significant part of the energy mix in the Nordic countries compared to Canada, Greenland, and Russia where a large number

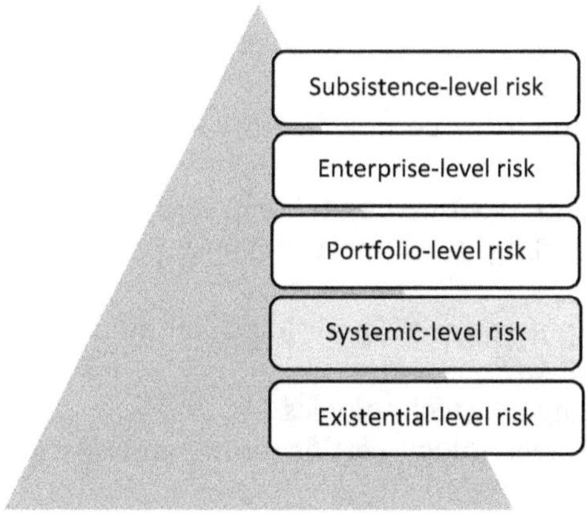

Figure 3.1 Risk hierarchy framework

The risk hierarchy framework shows different scales of risks. It starts with a subsistence level, meaning the risk to the livelihood of individuals and families thriving on natural resources such as fish and wild animals. It then moves to risk relevant to enterprises. The risk to portfolios includes industries. Systemic risk includes impact on various interconnected systems, and finally the framework presents a risk to the existence of people and/or other species.

Source: Adapted from Johannsdottir and Cook (2019, p. 6)

of local communities are not connected to a central energy infrastructure (grid), thus relying on diesel-fuel generators. Similar challenges are seen in Alaska, although the state is a forerunner in deploying renewable energy in remote communities (Poelzer et al., 2016; Arruda, 2018). Furthermore, there are enormous opportunities for Arctic communities to invest in renewable energy. Addressing local energy needs is expected to result in multiple benefits, including business and economic opportunities, well-being, and human security, and it will reinforce national and regional emission goals (Poelzer et al., 2016; Arruda, 2018).

There are nonetheless significant challenges in renewable energy development such as lack of human capacity and financial capital, legal issues concerning local and tribal governments, as well as technical and technological challenges. It has been suggested that guidelines developed with the Arctic Council's support would help mitigate these issues. To encourage development of non-binding specific Arctic Council Renewable Energy Guidelines and addressing prominent barriers, a group of Fulbright Arctic Initiative Scholars suggested 13 recommendations, principles, and best practices, for consideration (Poelzer et al., 2016, pp. 1–2), namely to:

1. Engage local communities early on in the planning process.
2. Establish a transparent system for sharing information about subsidies and costs.

3. Assess existing information about energy policies, institutional structures, and subsidies.
4. Share and develop core competencies at the local level.
5. Enable opportunities for regional collaboration.
6. Exploit opportunities for supporting complementary industries.
7. Anticipate how power system failures are addressed.
8. Utilize early-stage investments (venture capital, e.g. seed funding, start-up funding, early growth funding) to support high-risk projects in early stages of development.
9. Exercise opportunities of equity ownership for local communities.
10. Elaborate access to capital at various scales.
11. Construct funds for promising new technologies.
12. Advocate renewable energy technology certification standards relevant for Arctic conditions.
13. Foster opportunities for exporting knowledge.

The suggested guidelines would not hinder states from agreeing, where appropriate, similar or stricter standards (Poelzer et al., 2016).

Energy justice is a concept that relates to energy policy, systems, production, consumption, security, climate change, and activism. This concept has been discussed in the Arctic context, particularly in terms of energy infrastructure development, taking into account justice claims made by different stakeholders: governments, businesses, and civil society (McCauley et al., 2016). Different justice claims can be made but people and homes must come first when making decisions about energy development projects, thus requiring the involvement of locals in decisions that will affect their livelihoods. The underlying factor is that energy justice "aims to provide all individuals, across all areas, with safe, affordable and sustainable energy" (McCauley et al., 2016, p. 146). The questions raised in terms of energy injustice are in terms of distribution, recognition, and procedures (Jenkins et al., 2016): *Where are the injustices? Who is ignored? Is there fair process?* These questions also highlight energy access, the rights of local communities, and the role and obligation of governments.

Arctic communities also need to be seen as energy consumers, but in remote and sparsely populated areas communities struggle with rising costs of imported energy, such as diesel fuel, which is often the primary energy source for heating homes. To solve these issues, there is growing trend in local communities towards developing locally produced alternative energy sources (Arctic Council Sustainable Development Working Group, 2009), including off-grid energy systems and smart energy solutions (Len, 2016). To benefit from locally available resources "pooling relevant expertise from within and outside the Arctic to share experiences and exchange knowledge on sustainable energy challenges and solutions is one way to tackle common challenges". This can be applied across the region as well as building on local capacity and support systems (Len, 2016, p. 1). Sharing of knowledge can be done through networks, such as the Arctic Remote Energy Networks Academy (ARENA), endorsed by the Arctic Council's Sustainable Development Working Group (SDWG). ARENA is an example of a circumpolar platform for sharing knowledge through visits to community energy projects in different countries, and through knowledge exchanges, hands-on learning experiences, and mentorship (ARENA, 2020).

Adapting to change

Fundamental changes across the Arctic are experienced as a result of climate change. Socio-economic changes include demand for global resources, and increased tourism, transportation, and fisheries. These changes are often interlinked and interrelated and have impacts on people and ecosystems, consisting of "impacts on food and water security, infrastructure, and the goods and services provided by the area's ecosystems" (AMAP, 2017a, p. 13). Negative environmental and socio-economic impacts may undermine well-being, reduce food security, degrade infrastructure, and cause relocation of people. Diseases and increased exposure to pollutants may expand northward and put pressure on health services, as well as cause conflicts over land. These conflicts could intensify, given the opposing interests of economic activities, including extraction of natural resources or energy development projects and traditional activities, such as reindeer herding (AMAP, 2017a). In response to change, the Adaptation Actions for a Changing Arctic (AACA) project was adopted by the Arctic Council. Three pilot regions were studied: the Bering/Chukchi/Beaufort Region, Barents Region, and Baffin Bay/Davis Strait Region. These regions "provide a diverse range of socioeconomic and ecological conditions, as well as to include as many Arctic Council nations as possible" (AMAP, 2017a, p. 4; AMAP, 2018; AMAP, 2020).

Studies have focused on climate change and impacts related to temperature rise, future precipitation, wind speed, snow cover, ice sheet and lake ice formations, sea-surface temperatures, changing sea-levels, ocean- and sea ice, tundra ecosystems, and permafrost thawing (AMAP, 2017a, b, c). In addition, these studies also focus on other types of changes, including economic, demographic, technological, and political changes and trends (AMAP, 2017a) that will have profound impacts on locals who rely on subsistence, mixed cash-subsistence, or market economy activities for their livelihoods (AMAP, 2017c; Vammen Larsen et al., 2019).

Within the countries and communities included in the studies, there are wide dissimilarities and different capacities for adaptation. The capacity for adaptation depends on factors such as economic and environmental diversity, social conditions, and networks and mobility of organizations. For these reasons, adaptation actions need to be considered holistically and understood in the context of other changes within the region (AMAP, 2017c). Adaptation needs to take into account climate change, socio-economic, political, and cultural changes, and has to be seen as a long-term process rather than as an end-result (AMAP, 2017a).

In the case of the Baffin Bay/Davis Strait Region study, seven adaptation themes were identified: living and non-living (mineral extraction) resources, education, human health and well-being, infrastructure, shipping and tourism (AMAP, 2017b, c). The importance of significant global mitigation efforts is also emphasized as emission reduction will increase the likelihood of successful adaptation in the Arctic region. The speed and magnitude of environmental and socio-economic changes will impact people's ability to adapt (AMAP, 2017b, c), and one of the issues here is that adaptation is still at an early stage in the region. Regional differences are recognized but so too are significant gaps in risks addressed and adaptation actions carried out (Ford et al., 2014). Therefore, there are questions about the "overall readiness of governing bodies and communities to develop, implement, and promote adaptation" (Ford et al., 2017, p. 85).

If adaptation is successful, it should lead to the resilience of social-ecological systems, and in this process traditional and scientific knowledge has to be integrated (AMAP, 2017a). There are, however, barriers to action, such as a lack of motivation and sense of urgency, at times driven by impacts on economic results; trade-offs between immediate priorities and longer-term benefits of adaptation; inadequate resources, knowledge, and guidance from regional and national authorities; unclear responsibilities and inadequate frameworks; and overlooking of traditional and local knowledge. As a solution, it is important to strengthen those governing adaptation processes to fill in existing knowledge gaps (AMAP, 2017a, c); improve education and health; strengthen stakeholder and community involvement to utilize local and traditional knowledge (AMAP, 2017b); establish constructive and effective partnerships; ensure food security and sustainable seafood industry; and strengthen mixed economies (AMAP, 2017c). The governance characteristics supporting adaptation include the capacity to respond in a timely manner to change; flexibility of institutions responding to expected and unexpected change; interaction across scales of governance, such as in the case of system failures; co-management and partnership with local communities; mixture of longer-term adaptation and short-term disaster risk management; and recognition of change in a broader context, consisting of cumulative impacts and manifold stressors (AMAP, 2017c). The key aspect is also to deviate from reactive emergency responses to more proactive ones, and from governmental funding that is scattered and fragmented to adaptation planning that is endurable, thorough, and proactive (AMAP, 2017c).

Prior to initiating major projects, such as development of extractive resources, both environmental and social impact assessments should be carried out, and adaptation and scenario planning reinforced. The development should be in accordance with best international practices and benefit agreements negotiated, taking into account socio-cultural aspects, such as education, local involvement, procurement, and connectedness (AMAP,

2017b, c). Ecosystem approaches should, furthermore, be integrated into management, in order to protect biodiversity in the region, to strengthen cooperation across sectors and beyond national borders within the region (AMAP, 2017b), and to improve infrastructure (AMAP, 2017c).

It is claimed that barriers and limits to adaptation are in general political, where hard choices have to be made based on government involvement and intervention (Löf, 2013). The lens through which to view adaptive capacity, supporting or hindering adaptation efforts, can involve studying factors such as education, wealth, age, health status, poverty, livelihood, and populations. The next step would be to recognize the barriers hindering adaptive capacity from transforming into adaptation efforts. These include financial, cognitive, informational, technological, administrative, and organizational factors. To overcome these hindering factors and strengthen preparedness for adaptation, efforts could be made to examine leadership, decision-making, institutions, funding, stakeholder engagement, and relevant science factors (Ford et al., 2017). A general five-step approach to adaptation has been proposed: 1) assessment of climate vulnerability and key actors; 2) adaptation planning; 3) adaptation priorities where Multi-Criteria Decision Analysis (MCDA) and the Network-Centric Approach (NCA) can be used; 4) implementation; 5) tracking, monitoring, and evaluation (Champalle et al., 2015). When a community-specific adaptation strategy is developed for Arctic circumpolar communities, both known and potential environmental threats must be assessed, a community review of the data should be carried out and community decision made, an adaptation plan developed, and key indicators monitored (Berner et al., 2016). An available tool and information hub for sharing knowledge on climate change adaptation in the region is the Arctic Adaptation Exchange (AAE). It offers information for public and private sector decision-makers, communities, and researchers following a five-step process for adaptation and resilience (U.S. Climate Resilience Toolkit, n.d). Other adaptation tools for practical adaptation practices include guidelines, protocols, processes, and techniques (AMAP, 2017c).

Arctic resilience

The results of an Arctic Council project, led by the Stockholm Environment Institute (SEI) and the Stockholm Resilience Centre (SRC), were published in a report titled *Arctic Resilience Report 2016.* The research was carried out in collaboration with the Arctic countries, Indigenous Peoples, and a number of Arctic scientific organizations to gain understanding of climate change in the context of Arctic social-ecological systems, meaning a blend of human and natural systems, policies, and responses to the change occurring and building of resilience. Resilience, according to the report, entails the ability to shield from, or adapt to, disasters and stresses by passing through or influencing change (Arctic Council, 2016).

Biophysical impacts of climate change in the Arctic are well known (IPCC, n.d., IPCC, 2014), as well as economic opportunities and risks (Emmerson and Lahn, 2012; Johannsdottir and Cook, 2015; Arruda and Krutkowski, 2017). However, the impacts on "Indigenous Peoples whose livelihoods are disappearing, or whose villages are becoming uninhabitable" are less known; thus the holistic view of climate change and its impacts is missing (Arctic Council, 2016, p. x; Johannsdottir and Cook, 2019; Vammen Larsen et al., 2019). The resilience of communities in the region often depends on Indigenous Knowledge, referring to "understandings, skills and philosophies developed by societies with long histories of interaction with their natural surroundings" (IPCC, n.d., p. 552), but is often not accounted for in scientific knowledge. By focusing on social-ecological systems, the approach would become more all-inclusive of interaction between systems and how to respond to stresses and shocks, and thus build resilience critical for sustainable development. The social-ecological approach views people as a fundamental and instrumental part of the natural environment (Arctic Council, 2016).

To address climate change challenges in the Arctic, actors on different levels, within and outside the region, need to be a part of collective debate, decision-making, and take shared responsibility. This is necessary given that the region is a homeland to many inhabitants, is rich in resources, and is important for the global climate system. The issue is that different actors may view their interests differently with different material outcomes, and interests may be conflicting such as in the case of oil and gas extraction, commercial fisheries, infrastructure projects, and a subsistence way of life (Arctic Council, 2016; Johannsdottir and Cook, 2019; Vammen Larsen et al., 2019).

Climate change in the Arctic region is, in some cases, so substantial or abrupt that functions of systems are being fundamentally altered. These changes are named tipping points, and if the changes are irreversible they are considered as regime shifts (Arctic Council, 2016). In total, 19 regime shifts in marine, freshwater, and terrestrial ecosystems have been recognized in the Arctic, which are mainly due to climate change. These include, for instance, Arctic sea ice loss, Greenland ice sheet collapse, thermohaline circulation of ocean water, fishery collapses, mobility of Arctic residents, and impacts on tundra (Peterson and Rocha, 2016). These types of regime shifts have impacts on other systems, availability of subsistence food resources, and the well-being and sense of belonging of Arctic residents (Arctic Council, 2016). It should, however, be noted that these changes and impacts vary across countries within the region. Climate change mitigation actions are therefore critical, not just within the region but globally. National and local government also need to take action to reduce the risk of regime shifts, such as by developing policies and plans fostering resilience, and by developing management systems for natural resources (Arctic Council, 2016).

Studying cases from across the Arctic helps to build understanding of ecological, social, and behavioural processes that can support or erode resilience in the region. Through such analysis, four common elements beneficial to resilience have been recognized: "1) the capacity for self-organization – that is, to make decisions and implement responses to change; 2) diversity of responses to change; 3) the ability to learn from and integrate diverse types of knowledge; and 4) capacity to navigate surprise and uncertainty" (Arctic Council, 2016, p. xiii). Nevertheless, this is an understudied area.

Actions need to be carried out on individual, community, organizational, and institutional levels. Institutions play a crucial role, as they can bring people together for decision-making and direct and facilitate activities and actions, such as defining general policy problems, accumulate essential knowledge, set guiding rules and norms, and allocate resources. The Arctic Council plays a critical role as a high-level circumpolar policy forum, both addressing social and environmental issues and common concerns. It is also crucial in accumulating knowledge; suggesting solutions to common problems and opportunities, such as through policy and research; and fostering collaboration across different decision-making levels within and outside the region. Other institutions, both formal and informal, play a key role in building resilience. Informal institutions include communities of practices or networks (Arctic Council, 2016).

Seven types of capital are seen of importance in terms of how to cultivate and reinforce resilience: "natural capital, social capital, human capital, infrastructure, financial capital, knowledge assets and cultural capital" (Arctic Council, 2016, p. xv). These types of capital are intertwined and complementary but different combinations and contexts need to be considered. Efforts remain limited regarding monitoring and measuring these types of capital, as well as developing relevant indicators. Flexibility at the institutional level, taking into account social and environmental factors, is also seen of importance so that resilience can be strengthened and put into practical action. Six key questions need to be asked when activities, programmes, practices and/or strategies are evaluated in terms of how they contribute to building resilience: "1) Are the goals clear? 2) Are multiple kinds of knowledge being integrated? 3) Are place-based community partnerships being supported? 4) Are linkages being made across scales? 5) Is social learning being facilitated? 6) Is culture being taken into account?" (Arctic Council, 2016, p. xv).

Additionally, the status of social-ecological systems and how they are altered need to be monitored using a whole-systems perspective. Regime shifts need to be traced and studied; resilience assessment and modelling carried out; participatory situations in terms of knowledge, opinions, and values analysed; and tools for exploring issues collaboratively utilized. It is also seen of value to develop resilience strategies at regional and global levels, and solutions need be customized according to the specific context and conditions, taking both top-down and bottom-up approaches. What is, however, perhaps the most important aspect to building resilience in the Arctic is that "the people of the Arctic need support to organize, define challenges in their own terms, and find their own solutions, knowing that they will have the flexibility and external backing to implement their plans" (Arctic Council, 2016, p. xvi), as actions to build resilience will primarily be carried out within the region (Arctic Council, 2017). The challenge is that although there is promising resilience efforts underway, there are limiting factors which need to be mitigated, such as through sharing ideas, replicating, scaling up and adapting efforts that are promising, and strengthening coordination across different decision levels (Arctic Council, 2017).

Three major international agreements from 2015 draw attention to the characteristics of resilience: the Paris Agreement, the Sustainable Development Goals (SDGs), and the Sendai Framework for Disaster Risk Reduction (Arctic Council, 2016), all of which are of relevance for the Arctic people and Arctic residents, in particular Indigenous Peoples.

The Paris Agreement

The central aim of the 2015 Paris Agreement is to reinforce the global response to the climate change threat by limiting the global temperature rise in the 21st century to well below 2 degrees Celsius above pre-industrial levels, to carry out actions to limit the temperature increase even further to 1.5 degrees Celsius, and to strengthen countries' ability to deal with climate change impacts (UNFCCC, n.d.). The Paris Agreement was made within the United Nations Framework Convention on Climate Change (UNFCCC), and it stresses in Article 2b the "ability to adapt to the adverse impacts of climate change and foster climate resilience and low greenhouse gas emissions development, in a manner that does not threaten food production" (United Nations, 2015, p. 3). It furthermore puts forward a global goal of strengthening capacity for adaptation and resilience, and thus reducing vulnerability to climate change while contributing to sustainable development. The resilience concerns both socio-economic, i.e. communities and livelihood, and ecological systems, taking into account sustainable management of resources and economic diversification. In order to improve resilience to climate change and to reduce greenhouse gas emissions, there should be a long-term vison on how to enhance, utilize, and transfer technical development (Johannsdottir et al., 2014; United Nations, 2015).

The Paris Agreement furthermore acknowledges climate change as a common threat to the human race, and when actions are taken they should:

> respect, promote and consider their respective obligations on human rights, the right to health, the rights of indigenous peoples, local communities, migrants, children, persons with disabilities and people in vulnerable situations and the right to development, as well as gender equality, empowerment of women and intergenerational equity.
>
> (United Nations, 2015, p. 2)

Furthermore, traditional and Indigenous People's knowledge and knowledge systems should be taken into account and integrated, where appropriate, into relevant socio-economic and environmental policies and actions (United Nations, 2015).

The Sustainable Development Goals

The Sustainable Development Goals (SDGs) were adopted by United Nations Member States in 2015. At the core of the framework are 17 SDGs outlining how to ensure peace and prosperity of people and the planet now and into the future. To address poverty and other global issues, strategies need to be implemented that improve education and health, reduce inequality, and encourage economic growth. At the same time, the SDGs and strategies need to tackle climate change and preserve the oceans and forests (United Nations, n.d.a).

The 17 SDGs contain 169 targets and 231 unique indicators. There are 247 indicators in total, but twelve of them are repeated under particular targets (United Nations, n.d.c). None of these goals or targets mentions the Arctic

Figure 3.2 The UN Sustainable Development Goals

The figure shows the 17 interlinked Sustainable Development Goals adopted by all United Nations Member States in 2015. These Goals must be reached no later than 2030 if a sustainable future for all is to be ensured. A sustainable future considers peace, prosperity, people, and the planet, and should be ensured by global partnership between developed and developing countries. It requires action by governments, industries, the educational system, non-governmental organizations, and other stakeholders.

Source: United Nations (n.d.b)

in particular, but under Goal 2, Zero hunger, target 2.3 and indicator 2.3.2 specifically stress the importance of doubling the agricultural productivity and average incomes of small-scale food producers by 2030, such as in the case of Indigenous Peoples. This should be done, for instance, through equal and secure access to land, productive resources and inputs, knowledge, financial services, and markets (United Nations, n.d.c). Goal 4, Quality education, includes target 4.5 and indicator 4.5.1 aiming for eliminating gender disparities in education. This means that there should be equal access to all levels of education and vocational training for those in vulnerable situations, including Indigenous Peoples (United Nations, n.d.c; Arruda, 2019).

The SDGs also take into account resilience with regard to Goal 1, No poverty and target 1.5, where the aim is to reduce the exposure of the poor and vulnerable to climate-related extreme events and build their resilience so that they can cope with these events and other shocks and disasters of an economic, social, and environmental nature (United Nations, n.d.c). Goal 11, Sustainable cities and communities, also addresses resilience. According to target 11.b, there should be policies and plans in place towards mitigation and adaptation to climate change, resilience to disasters, inclusion, and resource efficiency, and the plans and policies should be in accordance with the Sendai Framework for Disaster Risk Reduction (United Nations, n.d.c). Goal 13, Climate actions, furthermore has a specific resilience target, 13.1, where the aim is to strengthen resilience and adaptive capacity to climate-related hazards and natural disasters in all countries. Goal 14, Life below water, additionally deals with resilience. In this case, strengthening of resilience is done by protecting marine and coastal ecosystems through sustainable management, so that significant adverse impacts can be avoided. Actions to restore marine and coastal ecosystems should furthermore be taken, so that healthy and productive oceans can be achieved.

The Sendai Framework for Disaster Risk Reduction

The Sendai Framework for Disaster Risk Reduction (Sendai Framework in short) is a voluntary non-binding agreement. It presents a global roadmap focusing on risk reduction in the period of 2015–2030. It was approved by the United Nations (UN) General Assembly after the 2015 third UN World Conference on Disaster Risk Reduction (WCDRR) (UNISDR, 2020). The Sendai Framework aims at reducing human and economic loss caused directly by disasters. It includes four priorities, seven targets for actions, and 13 guiding principles, with the aim of preventing new and reducing existing risk of disasters caused by natural or manmade hazards. Table 3.1 shows the priorities for actions and the targets of the Sendai Framework. Among the guiding principles is the primary role of States in preventing and reducing disaster risk, including through cooperation. The shared responsibility between central government and national authorities, sectors, and stakeholders as appropriate to national circumstances is also recognized, as well as protection

Table 3.1 Sendai Framework's priorities for actions and targets

Priorities for actions	*Targets*
1. Understanding disaster risk	1. Substantially reduce global disaster mortality by 2030, aiming to lower average per 100,000 global mortality between 2020 and 2030 compared to 2005–2015
1. Strengthening disaster risk governance to manage disaster risk	2. Substantially reduce the number of affected people globally by 2030, aiming to lower the average global figure per 100,000 between 2020 and 2030 compared to 2005–2015
3. Investing in disaster risk reduction for resilience	2. Reduce direct disaster economic loss in relation to global gross domestic product (GDP) by 2030
3. Enhancing disaster preparedness for effective response, and to 'Build Back Better' in recovery, rehabilitation, and reconstruction	4. Substantially reduce disaster damage to critical infrastructure and disruption of basic services, among them health and educational facilities, including through developing their resilience by 2030
	4. Substantially increase the number of countries with national and local disaster risk reduction strategies by 2020
	5. Substantially enhance international cooperation to developing countries through adequate and sustainable support to complement their national actions for implementation of this framework by 2030
	6. Substantially increase the availability of and access to multi-hazard early-warning systems and disaster risk information and assessments to people by 2030

Source: UNISDR (n.d.)

of persons and their assets, while all human rights, including the right to development, are promoted and protected (UNISDR, n.d.).

The Sendai Framework aims for a specific outcome:

> The substantial reduction of disaster risk and losses in lives, livelihoods and health and in the economic, physical, social, cultural and environmental assets of persons, businesses, communities and countries.
>
> (UNISDR, n.d.)

According to the Sendai Framework, governments ought to engage with relevant stakeholders in designing and implementing policies, plans, and standards (UNISDR, 2015). Of relevance in the Arctic region, and at national and local levels, is engagement with Indigenous Peoples regarding their knowledge of sensing early-warnings of risk, and to:

> ensure the use of traditional, indigenous and local knowledge and practices, as appropriate, to complement scientific knowledge in disaster risk assessment and the development and implementation of policies, strategies, plans and programmes of specific sectors, with a cross-sectoral approach, which should be tailored to localities and to the context.
>
> (UNISDR, 2015, p. 15)

CSR and the role of businesses and financial institutions in the Arctic

The Kiruna Declaration of the Arctic Council encourages enterprises operating in the Arctic to respect international principles and guidelines as a part of their work on CSR and sustainable business agendas, thus improving the local economic and social conditions (Arctic Council, 2013b). The dialogue has further advanced, such as in a workshop organized by the Arctic Athabaskan Council, one of six permanent participants to the Arctic Council at the Senior Arctic Officials meeting in Whitehorse, Canada in 2013. In a summary report from the workshop, CSR is seen as a fundamental human endeavour, raising the voice of local communities about the role of business to work and consult with them in ensuring sustainable development of the region, and to respect their values regarding sustainable development. CSR, furthermore, means that businesses should be responsible for the negative socio-economic-ecological impacts of their activities, support ecosystem management directed by local communities, and involve local voices in decision-making processes by states and corporations where investment decisions are made about the future of local communities. Short-term economic growth is viewed negatively, as well as overreliance on single industry, if it results in longer-term negative socio-ecological consequences (Arctic Athabaskan Council, 2013). The risk of greenwashing, where environmental claims are misleading or overstated, is also emphasized such as in the case of large industrial projects. There should be at minimum measurable net benefits to communities, where tangible improvements are made in terms of education or access to clean water and energy. Business development is also preferable on a smaller scale, such as through small and medium-sized businesses (Arctic Athabaskan Council, 2013).

CSR is mentioned in the Arctic Council's reports (AMAP, 2017d, p. 3); it is stated that for businesses' CSR, participatory management, increased transparency, digital economy, and infrastructure are of importance. So too is adaptation and adaptive management, applied in production and infrastructures situated in areas such as on the continental shelf, where there is permafrost, or in coastal zones. The same applies for technologies at risk in the case of extreme events. Although this is a problem-solving approach, it may not be viewed as adaptation or CSR, although in other cases factors such as addressing technological challenges, safety, and environmental concerns are integrated into CSR and corporate sustainable development programmes (Nikitina et al., 2015). In the case of perspectives from the Bering-Chukchi-Beaufort Region, Indigenous People have emphasized CSR in relation to mining companies and their programmes and how companies can improve the living conditions of locals and financially support cultural events. Given these outcomes, they believe that mining will benefit their district. However, issues are also recognized such as potential conflicts among Indigenous groups "over access to government funding, control over resource- and land-use projects, and over affiliations with corporate social responsibility (CSR) programs", and also that there is increasing risk of corruption (AMAP, 2017e, p. 228).

CSR has been discussed in the context of some of the large industrial projects taking place in the Arctic. These include the Arctic energy development projects in Russia where it is stated that "Russia's commitment to sustainability and corporate social responsibility (CSR) might become important for all Arctic countries as the possible negative impacts of its economic activity influence the whole territory of the far North" (Andreassen, 2016, p. 78). Local perceptions of petroleum developments in Hammerfest, Norway have also been studied, where local support is favourable given job opportunities, economic ripple effects of importance for the survival of the community, and attractiveness of the community in terms of livelihood and work. Lesser concerns were placed on environmental issues (Loe and Kelman, 2016). What has also been noticed is that if economic benefits are positive, the tolerance for environmental risk may be greater, and that both government and industry may fall short in supporting an equitable distribution of economic benefits. Thus, those that are closer to a petroleum industry, benefitting from their actions, tend to be more positive than others who are less involved (Kelman et al., 2016).

So-called social license to operate (SL or SLO), meaning an intangible license of locals supporting the operation, has also been discussed in the context of mining, where this intangible support is becoming increasingly important as the mining operations have negative environmental and social impacts which may outweigh economic benefits. These can be impacts on livelihoods, reindeer herding, or damage caused to the natural environment (Heikkinen et al., 2016). There may also be conflicts of interest between groups within a society, such if interests of reindeer herders do not coincide with interests of other groups in the case of petroleum activities (Loe et al., 2017). Transparency is therefore viewed of particular importance in mining (Heikkinen et al., 2016), and can also be assumed in the case of other large-scale industrial projects. CSR is also of importance as it is a common practice for mining companies operating in the region to follow some international standards, guidelines, or frameworks to deal with responsibility concerning sustainable development. CSR is regarded as companies' 'soft law' instruments, employed in addition to countries' 'hard law' instruments used to govern the mining industry. The general legal framework for mining companies is also supported by instruments such as environmental policies and regulations, environmental impact assessment (EIA), and land-use planning systems (Tiainen et al., 2015). What needs to be accounted for is the possible gap between CSR words and deeds, meaning that CSR emphasized by the company headquarters from a long distance away many not be the same as that practically carried out on the ground, such as in the case of oil spills and oil spill response (Henry et al., 2016).

CSR is also of relevance for the Arctic shipping industry where the subject of good ocean governance and stakeholders are relevant aspects of CSR. In some cases, this discussion is limited to environmental protection but excludes human elements such as safety of humans. Nevertheless, it is acknowledged that businesses are accountable to a range of stakeholders, which may have legal or moral claims due to business activities that may affect them. It is also recognized that CSR in shipping is less monitored than land-based sectors. For the shipping industry, the International Organization for Standardizations ISO 14000 Environmental Management System (EMS) is assumed as the most well known (Johannsson and Donner, 2015), as well as the more recent International Code for Ships Operating in Polar Waters (Polar Code) of the International Maritime Organization (IMO, 2017).

The role of insurance is emphasized in international agreements and frameworks, including the Paris Agreement and the Sendai Framework. The Paris Agreement specifies in Article 2c the role of finance, both in terms of pathways towards low greenhouse gas emissions and climate-resilient development. The Paris Agreement furthermore states the role of developed countries in providing climate finance to support developing countries in terms of mitigation, adaptation, and capacity building, e.g. education, training, and raising of public awareness. Insurance facilities, pooling of climate risk, and other insurance solutions are also mentioned in regard to cooperation and facilitation (United Nations, 2015).

Finance is also an important aspect of the SDGs. This is evident in Goal 1, No poverty, where the role of financial services and microfinance are seen of importance in ensuring men and women have equal rights to economic resources, but this is also important in regard to ownership and control over land and other forms of property, inheritance, natural resources, and appropriate new technology (United Nations, n.d.c). Goal 15, Life on land, and targets 15a and 15b, and indicator 15.1 all state the importance of financial resources used to benefit biodiversity, ecosystems, and forest management. Goal number 17, Partnership for the goals, and its targets and indicators all stress, one way or the other, the role of finance, whether through mobilizing financial resources, domestic taxes, gross national income (GNI), or foreign direct investment. The role of insurance is specifically addressed in Goal 8, Decent work and economic growth, target 8.10, where the aim is to strengthen the capacity of domestic financial institutions in order to encourage and expand access to banking, insurance, and financial services for all (United Nations, n.d.c).

In the case of the Sendai Framework (see Table 3.1), there was widespread acceptance and support from the insurance industry and intergovernmental organizations in signalling the willingness, capacity, support, and valuable role of the sector in supporting governments in disaster resilience. The economic argument was also at the forefront, in the sense that it is seen more cost-effective to invest in disaster risk reduction than to provide capital for response and recovery. Insurance can provide support in both cases, thus leading to less reliance on aid which is subject to the kindness of donors (UNISDR, 2015).

Further resources

A group of Fulbright Arctic Initiative scholars from Canada, Denmark, Sweden, and the United States proposed a recommendation for Arctic policymakers where they suggest shifting the focus from resilience, meaning coping with adverse situations, to thriving in terms of the health and well-being of communities. Four themes of importance for thriving communities, each with subthemes, are suggested (Cueva et al., 2019); see Table 3.2. The group also presented an Iceberg Model, where current monitoring and assessment is at the top of the iceberg, but there is a need to dive deeper under the surface to promote culture of health in Arctic communities and to raise questions such as: What are we looking for? What are we finding? Who decides what we are looking for? Under the surface are, for instance, community and regional contexts, community-driven research approaches, and in-depth understanding and knowledge about determinants of well-being (Cueva et al., 2019).

This approach to change the mindset from resilient to thriving raises questions about whether this approach is more appropriate than the mindset of merely coping, which appears in other cases involving issues and solutions, both within and outside the Arctic.

Table 3.2 Thriving communities: Themes and subthemes

Themes	*Subthemes*
1. Indigenous rights and knowledge	Indigenous rights, knowledge, Arctic Council and funding
2. Indigenous determinants of health	Local innovation, Indigenous leadership, informing systems, and under-represented groups
3. Monitoring and assessment	Monitoring, assessment, community-driven strategy, and information sharing
4. Community-led research approaches	Partnership, funding community research, reciprocity, and ethics

Source: Cueva et al. (2019)

Case study: Decline in investors' interest in Arctic oil projects and the risk of stranded assets

In an encyclical of Pope Francis on the environment he states the grave implications of environmental, social, economic, and political factors and for the distribution of goods, and how climate change is one of the challenges humanity is faced with today. Furthermore, he expresses his concerns that the worst impacts will be felt by the poor in developing countries relying on subsistence resources and ecosystem services for their livelihoods, given that they do not have the financial means to adapt to climate change impacts and natural disasters and therefore might have to migrate, given the negative impacts on animals and plants they rely on. The future of the poor and their children is therefore unsecure. The Pope, additionally, criticizes the current production and consumption models, and calls for policies that enable a drastic reduction of polluting gases such as carbon dioxide. He furthermore highlights the importance of access to clean and affordable renewable energy, the importance of developing new technologies, such as energy storage technologies, and the role of investment in solving these issues. The ideology of short-term profits is also criticized, while investment and other financial means are emphasized as part of solving urgent issues (Pope Francis, 2015).

Investors' decisions are currently influenced by scientific knowledge (e.g. the IPCC's synthesized reports), pressure from various stakeholders (e.g. Pope Francis' encyclical), ethical and moral dilemmas, risk of reputational damage, and risk of stranded assets. Recent development suggests a decline in investors' interest in Arctic oil projects. Major banks both in North America and Europe, including Barclays, Goldman Sachs, JPMorgan Chase & Co, and Wells Fargo and Co, are now refusing to finance any new upstream Arctic oil projects, either exploration and/or development, on environmental and social grounds (Arctic Today, 2019, 2020 a, b). Goldman Sachs (2019) refers to its Environmental Policy Framework, where climate mitigation, climate adaptation, climate risk management, and the bank's climate approach in its operation serve as the core, but also highlights the importance of the subsistence livelihoods of groups of Indigenous Peoples that have lived in the region for centuries. Other banks, including the Lloyds Banking Group (2018), have similar statements, policies, or frameworks, where they claim not to support project-specific finance which involves oil or gas exploration and production operations in the Polar regions.

Stranded assets often relate to environmental disturbance, for instance physical climate change impacts, resulting in consequent societal responses (Caldecott, 2017). Reasons for change are societal norms, legislation, disruptive innovation or advancement of the business context, or market forces (Bos and Gupta, 2019), both domestically or internationally. Subsequently, assets become non-performing or obsolete but still remain on companies' balance sheets as a loss of profit (Deloitte, 2020). These assets may have become non-performing or obsolete, and thus not generating expected revenues, due to unexpected or untimely deductions, devaluations, or conversion to liabilities (Caldecott et al., 2013). Often, stranded assets are associated with energy production such as coal, oil and gas, or nuclear production, low-carbon transition, and potential regulatory changes. Consequently, if assets are stranded so too are the resources; thus, the concept is related to the concept of stranded resources and the financial burdens or economic losses of investors. It has therefore been suggested that latecomers into an industry "need to be sensible in deciding which resources to develop to avoid carbon lock-in and whether phasing-in renewables could avoid creating stranded assets in the first place" (Bos and Gupta, 2019, p. 1). The International Renewable Energy Agency (IRENA) claims that stranding of assets will to some degree happen due to the low-carbon economy transition, and that it will impact buildings, upstream energy production, and power generation, followed by industry and industrial processes, mainly within the United States, Europe, Japan, China, and the Russian Federation. However, the IRENA report also states that if actions to deal with climate change are postponed due to delay in policy actions, then stranding of assets may become considerably more severe (IRENA, 2017). Divestment of asset owners and investors from coal, and in some cases oil and gas, are taking place in some cases due to ethical reasons, public concerns, or concerns over potential political actions. As a consequence, a large amount of oil and gas reserves may become stranded as the climate change risk escalates (Koehler and Bertocci, 2016).

Questions for discussion

1. What are the pros and cons of investors' decline in interest in Arctic energy investments?
2. What are the reasons for investors' decision to shy away from Arctic energy investments?
3. Who will be most affected by investors' decision to shy away from Arctic energy investments?
4. What role can finance play in building a resilient and sustainable future in the Arctic?
5. Does CSR play a role in building a resilient and sustainable future in the Arctic? If so, what role does it have?

Chapter summary

The key issues discussed in this chapter were climate-related concepts and region-specific climate change impacts, in terms of challenges and opportunities; Arctic energy development, infrastructure, transportation, and energy utilization, in terms of non-renewable and renewable energy development; the introduction of a risk-scale framework of importance when Arctic industrial projects are considered and carried out; and trade-offs between economic activities. Furthermore, a proposal for non-binding specific Arctic Council

Renewable Energy Guidelines was presented. Climate change adaptation was then discussed in the context of ecological and socio-economic changes, barriers and limits to adaptation, and adaptive capacity and approaches. Building and reinforcing resilience was then examined, such as through different types of capital, key actors, and international agreements emphasizing characteristics of resilience: the Paris Agreement, the Sustainable Development Goals, and the Sendai Framework for Disaster Risk Reduction. The role of businesses and financial institutions was discussed in the context of contemporary CSR trends in a world in transition. The idea of moving the discussion from building system resilience to thriving communities was also investigated. The case examined in this chapter was the decline in investors' interest in Arctic oil projects and the risk of stranded assets. This development is being influenced by factors such as scientific knowledge, pressure from various stakeholders (such as brought forth in Pope Francis' encyclical), ethical and moral dilemmas, and the risk of reputational damage to businesses.

References

Andreassen, N. (2016). Arctic energy development in Russia—How "sustainability" can fit? *Energy Research & Social Science*, 16, 78–88. http://dx.doi.org/10.1016/j.erss.2016.03.015

Anisimov, O.A., Vaughan, D.G., Callaghan, T.V., Furgal, C., Marchant, H., Prowse, T.D., Vilhjálmsson H., and Walsh, J.E. (2007). Polar regions (Arctic and Antarctic). In M.L. Parry, O.F. Canziani, J.P. Palutikof, P.J. van der Linden, and C.E. Hanson (Eds.), *Climate Change 2007: Impacts, Adaptation and Vulnerability. Contribution of Working Group II to the Fourth Assessment Report of the Intergovernmental Panel on Climate Change* (pp. 653–685). Cambridge, UK: Cambridge University Press.

Arctic Athabaskan Council. (2013). *Call and Response: Corporate Social Responsibility. The Key to Sustainable Communities.* Whitehorse, Canada: Author.

Arctic Council. (2013a). Glossary of terms. In *Arctic Resilience Interim Report 2013*. Stockholm, Sweden: Stockholm Environment Institute and Stockholm Resilience Centre.

Arctic Council. (2013b). Kiruna Declaration. Kriuna, Sweden: Author. Accessed on 3 April 2021 from https://oaarchive.arctic-council.org/handle/11374/93

Arctic Council. (2016). *Arctic Resilience Report*. M. Carson and G. Peterson (Eds.). Stockholm: Stockholm Environment Institute and Stockholm Resilience Centre. Accessed on 3 April 2021 from https://oaarchive.arctic-council.org/handle/11374/1838

Arctic Council. (2017). *Arctic Resilience Assessment. Synthesis for Arctic Leaders 2017*. Stockholm: Stockholm Resilience Centre, Stockholm University, and Stockholm Environment Institute. Accessed on 3 April 2021 from https://mediamanager.sei.org/documents/Publications/SEI-2017-Arctic-Resilience-Synthesis.pdf

Arctic Council Sustainable Development Working Group. (2009). *SDWG Report on Arctic Energy*. Ottawa, Canada: Author. Accessed on 3 April 2021 from http://library.arcticportal.org/1531/1/SDWG_ArcticEnergyReport_2009.pdf

Arctic Monitoring and Assessment Programme (AMAP). (2017a). *Adaptation Actions for a Changing Arctic (AACA): Barents Area Overview Report*. Oslo, Norway: Author.

Arctic Monitoring and Assessment Programme (AMAP). (2017b). *Adaptation Actions for a Changing Arctic (AACA): Baffin Bay/Davis Strait Region Overview Report.* Oslo, Norway: Author.

Arctic Monitoring and Assessment Programme (AMAP). (2017c). *Adaptation Actions for a Changing Arctic (AACA): Bering/Chukchi/Beaufort Region Overview Report.* Oslo, Norway: Author.

Arctic Monitoring and Assessment Programme (AMAP). (2017d). *Adaptation Actions for a Changing Arctic: Perspectives from the Barents Area.* Oslo, Norway: Author.

Arctic Monitoring and Assessment Programme (AMAP). (2017e). *Adaptation Actions for a Changing Arctic: Perspectives from the Bering-Chukchi-Beaufort Region.* Oslo, Norway: Author.

Arctic Monitoring and Assessment Programme (AMAP). (2018). *Adaptation Actions for a Changing Arctic: Perspectives from the Baffin Bay/Davis Strait Region.* Oslo, Norway: Author.

Arctic Monitoring and Assessment Programme (AMAP). (2020). Adaptation actions for a changing Arctic. Accessed on 30 March 2020 from https://aaca.amap.no/

Arctic Remote Energy Networks Academy (ARENA). (2020). ARENA: Arctic Remote Energy Networks Academy. Accessed on 30 March 2020 from http://arena.alaska.edu/2020/

Arctic Today. (2019). Citing climate change, Goldman Sachs rules out new Arctic oil financing. Accessed on 23 March 2020 from www.arctictoday.com/acknowledging-scientific-consensus-goldman-sachs-rules-out-new-arctic-oil-financing/

Arctic Today. (2020a). JPMorgan Chase becomes the second major US bank to drop Arctic oil financing. Accessed on 23 March 2020 from www.arctictoday.com/jpmorgan-chase-becomes-the-second-major-us-bank-to-drop-arctic-oil-financing/

Arctic Today. (2020b). Wall Street backs away from Arctic drilling amid Alaska political heat. Accessed on 23 March 2020 from www.arctictoday.com/wall-street-backs-away-from-arctic-drilling-amid-alaska-political-heat/

Arruda, G.M. (2014). Global governance, health systems and oil and gas exploration. *International Journal of Law and Management*, 56(6), 495–508.

Arruda, G.M. (2015). Arctic governance regime: The last frontier for hydrocarbons exploitation. *International Journal of Law and Management*, 57(5), 498–521.

Arruda, G.M. (2018). *Renewable Energy for the Arctic: New Perspectives.* Abingdon, UK: Routledge.

Arruda, G.M. (Ed.). (2019). *Renewable Energy for the Arctic: New Perspectives.* Abingdon, UK and New York: Routledge.

Arruda, G.M. and Krutkowski, S. (2017). Social impacts of climate change and resource development in the Arctic: Implications for Arctic governance. *Journal of Enterprising Communities: People and Places in the Global Economy*, 11(2), 277–288.

Berkes, F. and Folke, C. (1998). *Linking Social and Ecological Systems: Management Practices and Social Mechanisms for Building Resilience.* Cambridge, UK and New York: Cambridge University Press.

Berner, J., Brubaker, M., Revitch, B., Kreummel, E., Tcheripanoff, M., and Bell, J. (2016). Adaptation in Arctic circumpolar communities: Food and water security in a changing climate. *International Journal of Circumpolar Health* 75, 33820. http://dx.doi.org/10.3402/ijch.v75.33820

Bos, K. and Gupta, J. (2019). Stranded assets and stranded resources: Implications for climate change mitigation and global sustainable development. *Energy Research & Social Science*, 56, 101215. https://doi.org/10.1016/j.erss.2019.05.025

Caldecott, B. (2017). Introduction to special issue: Stranded assets and the environment. *Journal of Sustainable Finance & Investment*, 7(1), 1–13, doi:10.1080/20430795.2016.1266748

Caldecott, B., Howarth, N., and McSharry, P. (2013). *Stranded Assets in Agriculture: Protecting Value from Environment-Related Risks*. Oxford, UK: Smith School of Enterprise and the Environment, University of Oxford. Accessed on 3 April 2021 from www.smithschool.ox.ac.uk/publications/reports/stranded-assets-agriculture-report-final.pdf

Champalle, C., Ford, J.D., and Sherman, M. (2015). Prioritizing climate change adaptations in Canadian Arctic communities. *Sustainability*, 7(7), 9268–9292. doi:10.3390/su7079268

Comiso, J.C. and Hall, D.K. (2014). Climate trends in the Arctic as observed from space. *WIREs Climate Change*, 5, 389–409. doi:10.1002/wcc.277

Cueva, K., Guistini, S., Healey, G., Kanayurak, N., Larsen, C., Lavoie, J., Rink, E., and Stoor, J.P. (2019). From resilient to thriving: Supporting health and well-being in Arctic communities. In *2019 Policy Brief: Fulbright Arctic Initiative Symposium*, Washington D.C.

Deloitte. (2020). *Clarity in Financial Reporting. Disclosure of Climate-Related Risks*. Accessed on 8 April 2020 from www2.deloitte.com/content/dam/Deloitte/au/Documents/audit/deloitte-au-audit-clarity-disclosure-climate-related-risks-070220.pdf

Emmerson, C. and Lahn, G. (2012). *Arctic Opening: Opportunity and Risk in the High North*. London: Lloyd's and Chatham House.

Ford, J.D., Labbé, J., Flynn, M., and Araos, M. (2017). Readiness for climate change adaptation in the Arctic: A case study from Nunavut, Canada. *Climate Change*, 145, 85–100. doi:10.1007/s10584-017-2071-4

Ford, J.D., McDowell, G., and Jones, J. (2014). The state of climate change adaptation in the Arctic. *Environmental Research Letters*, 9, 104005. doi:10.1088/1748-9326/9/10/104005

Gautier, D.L., Bird, K.J., Charpentier, R.R., Grantz, A., Houseknecht, D.W., Klett, T. R., Moore, T.E., Pitman, J.K., Schenk, C.J., Schuenemeyer, J.H., Sørensen, K., Tennyson, M.E., Valin, Z.C., and Wandrey, C. J. (2009). Assessment of undiscovered oil and gas in the Arctic. *Science*, 324(5931), 1175–1179. doi:10.1126/science.1169467

Goldman Sachs. (2019). *Goldman Sachs Environmental Policy Framework*. New York: Author. Accessed on 3 April 2021 from www.goldmansachs.com/citizenship/environmental-stewardship/epf-pdf.pdf

Royal Greenland (n.d.). About Royal Greenland. Accessed on 3 April 2021 from www.royalgreenland.com/royal-greenland/about-royal-greenland/

Griffin, P. (2017).*The Carbon Majors Database. CDP Carbon Majors Report 2017*. London: CDP. Accessed on 3 April 2021 from www.cdp.net/en/articles/media/new-report-shows-just-100-companies-are-source-of-over-70-of-emissions

Heikkinen, H., Lépy, É., Sarkki, S., and Komu, T. (2016). Challenges in acquiring a social licence to mine in the globalising Arctic. *Polar Record*, 52(4), 399–411. doi:10.1017/S0032247413000843

Henry, L.A., Nysten-Haarala, S., Tulaeva, S., and Tysiachniouk, M. (2016). Corporate social responsibility and the oil industry in the Russian Arctic: Global norms and neo-paternalism. *Europe-Asia Studies*, 68(8), 1340–1368. http://dx.doi.org/10.1080/09668136.2016.1233523

Intergovernmental Panel on Climate Change (IPCC). (n.d.). Special Report: Global Warming of 1.5 °C. Glossary. Accessed on 23 March 2020 from www.ipcc.ch/sr15/chapter/glossary/

Intergovernmental Panel on Climate Change (IPCC). (2014). Summary for policymakers. In C.B. Field, V.R. Barros, D.J. Dokken, K.J. Mach, M.D. Mastrandrea, T.E. Bilir, M. Chatterjee, K.L. Ebi, Y.O. Estrada, R.C. Genova, B. Girma, E.S. Kissel, A.N. Levy, S. MacCracken, P.R. Mastrandrea, and L.L.White (Eds.), *Climate Change 2014: Impacts, Adaptation, and Vulnerability. Part A: Global and Sectoral Aspects. Contribution of Working Group II to the Fifth Assessment Report of the Intergovernmental Panel on Climate Change* (pp. 1–32). Cambridge, UK and New York: Cambridge University Press.

International Maritime Organization (IMO). (2017). International Code for Ships Operating in Polar Waters (Polar Code). Accessed on 7 April 2020 from www.imo.org/en/MediaCentre/HotTopics/polar/Pages/default.aspx

International Renewable Energy Agency (IRENA). (2017). *Stranded Assets and Renewable: How the Energy Transition Affects the Value of Energy Reserves, Buildings and Capital Stock*. Abu Dhabi: Author. Accessed on 3 April 2021 from www.irena.org/-/media/Files/IRENA/Agency/Publication/2017/Jul/IRENA_REmap_Stranded_assets_and_renewables_2017.pdf

Jenkins, K., McCauley, D., Heffron, R. and Stephan, H., (2016). Energy justice: A conceptual review. *Energy Research & Social Science*, 11, 174–182. http://dx.doi.org/10.1016/j.erss.2015.10.0042214-6296

Johannsdottir, L. and Cook, D. (2015). *An Insurance Perspective on Arctic Opportunities and Risks: Hydrocarbon Exploration & Shipping. Reykjavík*. Working paper published by the Institute of International Affairs, the Centre for Arctic Policy Studies. Accessed on 2 April 2021 from https://doi.org/10.1016/j.erss.2015.10.004

Johannsdottir, L. and Cook, D. (2019). Systemic risk of maritime-related oil spills viewed from an Arctic and insurance perspective. *Ocean & Coastal Management*, 179(1): 104853. doi:10.1016/j.ocecoaman.2019.104853

Johannsdottir, L., Goodsite, M., Davidsdottir, B. and Olafsson, S. (2014). Insurers' role in enhancing development and utilization of environmentally sound technologies: A case study of Nordic insurers. *Journal of Cleaner Production*, 65, 526–538.

Johannsson, T. and Donner, P. (2015). *The Shipping Industry, Ocean Governance and Environmental Law in the Paradigm Shift. In Search of a Pragmatic Balance for the Arctic*. Cham, Heidelberg, New York, Dordrecht, London: Springer.

Kelman, I., Loe, J.S.P., Rowe, E.W., Wilson, E., Poussenkova, N., Nikitina, E., and Fjærtoft, D.B. (2016). Local perceptions of corporate social responsibility for Arctic petroleum in the Barents Region. *Arctic Review on Law and Politics*, 7(2), 152178. http://dx.doi.org/10.17585/arctic.v7.418 1

Koehler, D.A. and Bertocci, B. (2016). *Stranded Assets: What Lies Beneath*. Chicago, IL: UBS Asset Management (Americas) Inc.

Len, C. (2016). *Facilitating Remote Energy Access in the Arctic and Southeast Asia: Key Conditions and Initiatives* (Policy Brief no. 16). Singapore: Energy Studies Institute.

Lloyds Banking Group. (2018). Oil & Gas Sector Statement. Accessed on 30 March 30 2020 from www.banktrack.org/download/oil_gas_sector_statement/lbg_sector_statements_oilgasfinal20180128v5.pdf

Loe, J.S.P. and Kelman, I. (2016). Arctic petroleum's community impacts: Local perceptions from Hammerfest, Norway. *Energy Research & Social Science*, 16, 25–34. http://dx.doi.org/10.1016/j.erss.2016.03.008

Loe, J.S.P., Kelman, I., Fjærtoft, D.B., and Poussenkova, N. (2017). Arctic petroleum: Local CSR perceptions in the Nenets region of Russia. *Social Responsibility Journal*, 13(2), 307–322. https://doi.org/10.1108/SRJ-10-2015-0150

Löf, A. (2013). Examining limits and barriers to climate change adaptation in an Indigenous reindeer herding community. *Climate and Development*, 5(4), 328–339. doi.org/10.1080/17565529.2013.831338

Mack, M.C., Bret-Harte, M.S., Hollingsworth, T.K.N., Jandt, R.R., Schuur, E.A.G., Shaver, G.R., and Verbyla, D. L. (2011). Carbon loss from an unprecedented arctic tundra wildfire. *Nature*, 475, 489–492.

McCauley, D., Heffron, R., Pavlenko, M., Rehner, R., and Holmes, R. (2016). Energy justice in the Arctic: Implications for energy infrastructural development in the Arctic. *Energy Research & Social Science*, 16, 141–146. http://dx.doi.org/10.1016/j.erss.2016.03.0192214-6296

Nikitina, E., Poussenkova, N., Loe, J., and Kelman,I. (2015). Sustainable oil and gas practices in the Arctic: Business responsibility towards communities . *The Arctic Herald*, 3, 60–74.

Overland, J.E., Wang, M., Walsh, J.E., and Stroeve, J.C. (2013). Future Arctic climate changes: Adaptation and mitigation time scales. *Earth's Future*, 2, 68–74. doi:10.1002/2013EF000162

Peterson, G. and Rocha, J.C. (2016). Arctic regime shifts and resilience. In M. Carson and G. Peterson (Eds.), *Arctic Resilience Report*. Stockholm: Stockholm Environment Institute and Stockholm Resilience Centre. Accessed on 2 April 2021 from https://oaarchive.arctic-council.org/handle/11374/1838

Poelzer, G., Hoogensen Gjorv, G., Holdmann, G., Johnson, N., Magnússon, B.M., Sokka, L., Tysyachnyouk, M., and Yu, S. (2016). *Developing Renewable Energy in Arctic and Sub-Arctic Regions and Communities: Working Recommendations of the Fulbright Arctic Initiative Energy Group*. Saskatoon, Canada: University of Saskatchewan, International Centre for Northern Governance and Development.

Pope Francis. (2015). *Laudato Si': On Care for Our Common Home* [Encyclical]. Rome: Vatican.

Rees, A. and Sharp, D. (2011). *Drilling in Extreme Environments: Challenges and Implications for the Energy Insurance Industry in Extreme Environments*. London: Lloyd's of London.

The White House & Prime Minister of Canada. (2016). U.S.-Canada Joint Statement on Climate, Energy, and Arctic Leadership (Statements & Releases). Accessed on 26 March 2020 from https://obamawhitehouse.archives.gov/the-press-office/2016/03/10/us-canada-joint-statement-climate-energy-and-arctic-leadership and https://pm.gc.ca/en/news/statements/2016/03/10/us-canada-joint-statement-climate-energy-and-arctic-leadership

Tiainen, H., Sairinen, R., and Sidorenko, O. (2015). Governance of sustainable mining in Arctic countries: Finland, Sweden, Greenland & Russia. In L. Heininen (Ed.), *Arctic Yearbook 2015. Northern Research Forum and the University of the Arctic Thematic Network (TN) on Geopolitics and Security*. Accessed on 3 April 2021 from www.uarctic.org/news/2015/10/arctic-yearbook-2015-launching-october-16th/

United Nations. (n.d.a). Sustainable Development Goals Knowledge Platform. Accessed on 1 April 2020 from https://sustainabledevelopment.un.org/sdgs

United Nations. (n.d.b). Sustainable Development Goals. Accessed on 1 April 2020 from www.un.org/sustainabledevelopment/news/communications-material/

United Nations. (n.d.c). Global indicator framework for the Sustainable Development Goals and targets of the 2030 Agenda for Sustainable Development. Accessed on 1 April 2020 from https://unstats.un.org/sdgs/indicators/Global%20Indicator%20 Framework%20after%202020%20review_Eng.pdf

United Nations. (2015). Paris Agreement. Paris: Author. Accessed on 3 April 2021 from https://unfccc.int/sites/default/files/english_paris_agreement.pdf

United Nations Framework Convention on Climate Change (UNFCCC). (n.d.). The Paris Agreement. Accessed on 1 April 2020 from https://unfccc.int/process-and-meetings/the-paris-agreement/the-paris-agreement

United Nations International Strategy for Disaster Reduction (UNISDR). (n.d.). Chart of the Sendai Framework for Disaster Risk Reduction. Accessed on 1 April from www.preventionweb.net/files/44983_sendaiframeworkchart.pdf

United Nations International Strategy for Disaster Reduction (UNISDR). (2015). Sendai Framework for Disaster Risk Reduction 2015–2030. Geneva: United Nations Office for Disaster Risk Reduction.

United Nations International Strategy for Disaster Reduction (UNISDR). (2020). What is the Sendai Framework for Disaster Risk? Accessed on 1 April 2020 from www.undrr.org/implementing-sendai-framework/what-sendai-framework

U.S. Climate Resilience Toolkit. (n.d.). Arctic Adaptation Exchange. Accessed on 6 April 2020 from https://toolkit.climate.gov/tool/arctic-adaptation-exchange

Vammen Larsen, S., Bors, E., Johannsdottir, L., Gladun, E., Gritsenko, D., Nysten-Haarala, S., Tulaeva, S., and Sformo, T. (2019). A conceptual framework of Arctic economies for policy-making, research, and practice. *Global Policy*, 10(4): 686–696. doi:10.1111/1758-5899.12720

4 The eight Arctic nations and their respective industries

Case studies

Learning objectives

- Understand that the CSR emphasis differs depending on types of industries.
- Recognize that CSR is also relevant to controversial industries.
- Understand the importance of collaboration with stakeholders.
- Understand the importance of transparency for building and maintaining trust.
- Recognize that business founded on traditional culture and knowledge can grow into international business.
- Recognize that research projects may hold a key to solving global problems, such as climate change.

Introduction

> Eight countries have regions belonging to the Arctic economies: United States, Canada, Denmark, Iceland, Norway, Sweden, Finland and Russia. The Arctic regions Greenland and Faroe Islands are represented as sovereign countries corresponding to their extensive Self-Governance Rule within the realm of Denmark.
>
> (Mäenpää et al., 2017, p. 27)

While the Arctic covers 8% of the world's land area, it only has 10 million inhabitants. Approximately one million of the total Arctic population is Indigenous, from more than 40 distinctive ethnic groups, with the highest population (75%) in the Canadian Arctic and in Greenland (Nordregio, n.d).

There are relatively few common denominators for the Arctic nations: mainly the vast wilderness area and the importance of the ocean and what it provides in the form of resources, both living resources and minerals (Mäenpää et al., 2017). Other similarities include small economies, limited diversification of economic activities, and reliance on raw material exports (Glomsrød et al., 2017). There are many differences such as in terms of the size of the landmass, the population size and Indigenous Peoples, economic activities (Glomsrød et al., 2017; Mäenpää et al., 2017; Arruda, 2019). The industries are also heterogeneous, although some clusters are recognizable, including petroleum and mineral mining in Arctic Russia, Alaska, and northern Canada; natural resource extraction in Greenland and the Faroe Islands; and secondary industries in Arctic Finland and Sweden (Glomsrød et al., 2017). Traditional subsistence activities, including hunting and fishing, are also of great importance for livelihoods and lifestyles, such as in Greenland and northern Canada, and reindeer herding in Arctic Russia and Scandinavia (GRIDA, 2005).

Background: From theory to practice

This section focuses on business practices rather than theory. In this context, it is worth mentioning the Arctic Economic Council (AEC), which is an independent organization created by the Arctic Council during the Canadian chairmanship (2013–2015). Its inaugural meeting took place in Iqaluit, Nunavut on 2–3 September 2014. The role of the Arctic Economic Council is to "facilitate Arctic business-to-business activities and responsible economic development through the sharing of best practices, technological solutions, standards, and other information" (AEC, n.d.). The organization further states that it supports market accessibility and contributes to the work of the Arctic Council by providing a business perspective and advice. Various businesses operating in the region are AEC members, ranging from Indigenous economic development corporations reindeer herders to mining and shipping companies. The focus areas of AEC include infrastructure-related matters, such as aviation, maritime transportation, communication, IT, energy-related activities, mining, tourism, blue economy, and investment in human resources and capacity building (AEC, n.d.). Arctic business affairs are also discussed in forums such as the Arctic Business Forum conferences. In the 2019 conference report it is claimed that "the future of Arctic business depends on the Climate change", and that there are "problems created by the global warming". Therefore, business growth needs to be responsible and sustainable (Rautajoki and Lakkapää, 2019, p. 8).

In the following discussion, each Arctic nation is introduced briefly in terms of landmass size, population, specifically Arctic population and Indigenous Peoples, and their main industries and value-added activities. This is followed by a discussion about practical business-related cases from each of the Arctic nations. The purpose of the case studies is to provide practical examples of different cases. Operations of some companies introduced are controversial (e.g. hydrocarbons and mining), and perhaps not without reason such as in the case of toxic chemicals and metals (Alaska Community Action on Toxics, 2020; State of Alaska, 2019), but corporate social responsibility (CSR) is still of relevance for such companies. Whether companies' operations are controversial or not, it is critically important to be transparent both about positive and negative aspects of the operation as it is the foundation for establishing and maintaining trust, leads to continuous improvements, and is essential for restraints provided by authorities and stakeholders, such as NGOs and industry associations. For instance, the Carbon Disclosure Project (CDP) tracks the performance of companies in areas such as climate, forests, water, and supply chains. Not all companies requested to respond to the CDP do so, and if they fail to respond or provide insufficient information, their score will be F.

Table 4.1 lists examples of the Arctic nation's respective industries, and the title of each case. The cases covered range from fine art and culture to mining and hydrocarbons, from small newly founded companies to large industry groups and networks, and from cooperatives to government-owned and limited liability companies. This discussion is by no means exhaustive, as several key industries are the backbone of the economy for each nation or region. The remainder of this chapter is structured in such a way that key information about each country is first provided, followed by a discussion about a selected national cases. In the case of Denmark, the case discussed is Greenlandic, and other cases have strong relevance for the Arctic region.

Canada

Canada is geographically a part of northern North America. Politically, the Canadian Arctic refers to three territories: the Northwest Territories, Nunavut, and Yukon. The total area of Canada is 9,984,670 sq km, and around 40% of the landmass is considered Arctic and Northern (Arctic Council, n.d.a; Central Intelligence Agency, n.d.a). The total population is around 38 million inhabitants, with the Canadian Arctic being home to around 150,000 inhabitants. Of these more than half are Indigenous. Less than 1% of Canada's population, therefore, lives in the Arctic region (Arctic Council, n.d.a; Central Intelligence Agency, n.d.a). Three Permanent Participant organizations to the Arctic Council, with Canadian constituents, are the Arctic Athabaskan Council, the Inuit Circumpolar Council, and the Gwich'in Council International (Arctic Council, n.d.a).

The government of Canada has defined the nation's Arctic capacities by sectors. Indigenous traditional livelihoods, namely hunting, fishing, and arts and crafts, are identified as the long-lasting lifeline and mainstay of the local economy. Additionally, other economic activities are recognized: ice, ocean, and clean Arctic technologies and services; cold-climate manufacturing; northern infrastructure; engineering and design; transportation and logistics services; mining equipment and services; Arctic maritime, harsh environment oil and gas equipment and services; information communications technology; Arctic earth observation; aerospace and defense; northern cultural, creative, and specialty food industries; and Arctic research and education (Government of Canada, 2016). In terms of value added in selected industries, mining, quarrying, and oil and gas extraction; utility, construction, and manufacturing; government education and health services; and public administration

Table 4.1 The Arctic nations, their respective industries, and selected cases

Canada	*Denmark/ Greenland*	*Finland*	*Iceland*	*Norway*	*Sweden*	*The Russian Federation*	*The United States*
Fine art and culture	Fisheries	Mining	Renewable energy	Tourism	Adventure products	Natural gas and liquid hydrocarbons	Mining
West Baffin Eskimo Cooperative	The Royal Greenland Group	The Sustainable Mining Network	CarbFix	The Association of Arctic Expedition Cruise Operators	Tentipi	Novatek	Teck's Red Dog mine

and defence are important in all three Canadian Arctic territories (Glomsrød et al., 2017).

The West Baffin Eskimo Co-operative: A case study from Canada

The West Baffin Eskimo Co-operative, also called the Kinngait Co-operative, was established in 1959. It is an Inuit Co-operative and its members are Inuit artists who create exquisite internationally known carvings, drawings, and prints. The cooperative is a part of Arctic Co-operatives Limited, which is a "service federation that is owned and controlled by 32 community-based Co-operative businesses that are located in Nunavut, Northwest Territories and Yukon" (Arctic Co-operatives Limited, n.d). The West Baffin Eskimo Co-operative operation takes place in the Kinngait Studios in Cape Dorset on Dorset Island at the southern tip of Baffin Island in the Qikiqtaaluk Region of Nunavut (WBEC, n.d.b). In Inuktitut, Kinngait means mountain (WBEC, n.d.a). The cooperative also operates an office in Toronto: Dorset Fine Arts established in 1978 (DFA, n.d.). The role of Dorset Fine Arts is to interface with the art market locally, regionally, and globally, including galleries, museums, cultural professionals, and Inuit art enthusiasts (DFA, n.d.). Since its establishment, the role of the West Baffin Eskimo Co-operative has expanded significantly; currently, it includes "communications, promotion, advocacy, government relations and special projects as related to the Inuit art of Cape Dorset" (WBEC, n.d.b).

When artists in the Cape Dorset started to produce drawings and prints, it was on a small scale, but has since "developed into a major northern art community" (WBEC, n.d.a). This is due to the commitment of James and Alma Houston (1951 to 1962) and Terry Ryan who first had a role as an art advisor in 1960 before taking over as the manager of the West Baffin Eskimo Co-operative (1962–2001) (WBEC, n.d.a). Since the establishment of the West Baffin Eskimo Co-operative, generations of Cape Dorset artists have produced sculptures based on local traditions sold by the cooperative. Artists are paid "based on the originality, quality, and complexity of their sculpture" (WBEC, n.d.a). The traditional sculptures crafted by the locals are based on Arctic wildlife and nature, including caribou, dancing bears and magical birds, and Inuit culture, showing supernatural and spiritual forms where animal and human forms are merged (WBEC, n.d.a). Artists take pride in astonishingly thin elements of their sculptures, demonstrating their skills and the strength of the stone used for sculpturing. Given the quality of the artwork, over a "dozen artists from Cape Dorset have been made members of the Royal Canadian Academy of Arts" (WBEC, n.d.a).

Worldwide, the West Baffin Eskimo Co-operative is known for the originality and quality of limited-edition prints crafted by its artist members in the currently known Kinngait Studios. This studio is "the longest continuous running print studio in Canada" (WBEC, n.d.c). The success of the formative years of printmaking in Cape Dorset is attributed to an artist from Toronto,

James Houston, who stayed with his family in the community for ten years, from 1952 onward. The method used at the beginning was sealskin stencils, and then the usefulness of local carving stone for printmaking was discovered, followed by the lithography printing process. More recently, various methods began to be used, providing the artists with a variety of methods to develop the art collection (WBEC, n.d.c).

This case illustrates how nurturing local culture can create an income stream for local artists and their families through an international reach of exquisite Inuit artwork known for its originality and quality.

Denmark – Greenland

The Kingdom of Denmark includes Denmark, Greenland, and the Faroe Islands; Greenland and Faroe Islands are self-governing territories within the Kingdom. Greenland is the world's largest island and the home-rule government is responsible for most of the country's domestic affairs. Greenland is geographically a part of northern North America but geopolitically a part of Europe. The total area of Greenland is 2,166,086 sq km, and 80% is ice-capped. Greenland is located between the Arctic Ocean and the North Atlantic Ocean, northeast of Canada, and two-thirds of the island's landmass is within the Arctic Circle. The population of Greenland is estimated at around 57,600 inhabitants (Central Intelligence Agency, n.d.b). The Arctic Indigenous Peoples in Greenland are Intuits (Arctic Council, n.d.b).

The Greenlandic economy has four main pillars: "fisheries, raw materials, tourism and other land-based business" (OECD, 2005, p. 1). Of those, the primary industry is fishing (88%) and fisheries, where commercial stocks are Greenland halibut, shrimps, and cod, but growing and/or important industries include energy and minerals, tourism, agriculture, and hunting of seals, whales, and other mammals which are crucial for the survival of the Greenlanders (Government of Greenland, n.d.). In terms of value added by industry, unspecified service activities yield the highest value, followed by wholesale and retail trade, construction, public administration and defence, and transportation and storage (Glomsrød et al., 2017).

The Royal Greenland Group: A case study from Greenland

Royal Greenland A/S is an independent limited liability company. The Greenlandic government owns all of its shares. The company combines the latest modern technology with its heritage of long traditions, traditional fishing methods, and expertise of the local Greenlandic fishing communities. The key species the company catches are cold-water prawns, Greenland halibut, snow crab, cod, and lumpfish; the fishing, processing, and seafood trades are the foundation for the livelihood of the local Greenlandic communities (Royal Greenland, n.d.c). The company is the world's largest supplier of cold-water prawns and Greenland halibut, while at the same time it has a very strong local presence. In 2018, it bought fresh seafood from more than 1,000 independent local fishermen, and the company states that it is "committed to creating prosperity and development for the fishing communities that deliver our seafood" (Royal Greenland, n.d.a).

The company's mission and vision highlights that the value of marine resources should be sustainably maximized, products should be safe and healthy, earnings should be sustainable, the risk managed, and individuals respected. Royal Greenland also emphasizes integrity and transparency and makes the following planet promise: "We manage the resources to which we have access, and upon which our future depends, sustainably" (Royal Greenland, n.d.a). The company has a strong focus on sustainably, namely people-profit-planet, and states that it "has achieved significant results in reducing water and energy consumption, as well as within sustainable fishery" (n.d.b), claiming that social, environmental, and financial performance go hand in hand.

Royal Greenland has published both a CSR policy and a sustainability strategy. The people's perspective focuses on employee development, working conditions, and human rights. The planet perspective entails a focus on sustainable fisheries and utilization of resources while the profit aspects build on business ethics. The UN Global Compact's principles are the foundation for the company's sustainability strategy and two key principles guide its strategic approach (Royal Greenland, n.d.b):

1. Environmental and social development and initiatives must support economic growth
2. Business growth must support environmental and social development in the local community

The CSR and sustainability policies of Royal Greenland include a CSR policy, anti-corruption policy, modern slavery statement, climate and environmental policy, equal gender policy, sustainable fishery policy, human rights policy, and bullying and harassment policy. Furthermore, the company has published a code of conduct and a code of conduct for its suppliers (Royal Greenland, n.d.d). The company is a member of the Nordic Biopanel, whose aim is to work towards "developing a Nordic strategy for developing the bio-economic utilization of natural resources" (Royal Greenland, n.d.d). As an example of the bio-economic utilization of natural resources, the company has "utilized prawn shells from our prawn factory in Ilulissat, Greenland, making flour fit for human consumption, doubling the utilization rate" of the prawn resources (Royal Greenland, n.d.d).

As an example of sustainability collaboration, the company together with 20 other companies is taking part in a project supervised by the Confederation of Danish Industry. It is titled *From Philanthropy to Business*, and the "aim of the project is to inspire and motivate companies to develop new growth areas which can promote the achievement of the UN's 17 Global

Sustainable Development Goals towards 2030" (Royal Greenland, 2019). According to the company's 2018 annual report, the main focus is on Goal 4, quality education, Goal 8, decent work and economic growth, Goal 14, life below water, and Goal 12, responsible consumption and production (Royal Greenland, 2019).

The Royal Greenland case demonstrates how local government can run a successful business, taking into account both the latest modern technology and traditional fishing methods, thus sustainably utilizing marine resources while maximizing economic value for the benefit of society.

Finland

Finland is in northern Europe, between Sweden and Russia. The total area of Finland is 338,145 sq km, and one-third of the landmass, or territories, lies above the Arctic Circle in the provinces of Lapland (Arctic Council, n.d.c). The total population of Finland is close to 5.6 million inhabitants. The Finnish Arctic population totals 180,000 inhabitants, with the Saami population totalling approximately 10,500 inhabitants (Arctic Council, n.d.c; Central Intelligence Agency, n.d.c). The following provinces are Arctic: Northern Ostrobothnia, Kainuu, and Lapland; the Indigenous Peoples in the Arctic are Saami (Arctic Council, n.d.c).

The main industries in northern Finland are forestry, natural resources extraction, mining and metal industries, tourism, bio and renewable energy, offshore industries, and shipping (Arctic Institute, n.d.), in addition to construction, environmental and maritime technology, Arctic infrastructure development, navigation and transportation in icy waters, and research and education (Arctic Council, n.d.b). The biggest employers in Lapland are forestry, mining, and tourism (Arctic Institute, n.d.). In terms of value added by industry in the Arctic Finland, manufacturing of high-tech goods and electronics are recognized as of importance, in addition to health care and social work, real estate activities, education, and construction (Glomsrød et al., 2017).

The Sustainable Mining Network: A case study from Finland

The Sustainable Mining Network, Kaivosvastuu, established in 2014, is a neutral platform used for interaction between the mining industry, its stakeholders, and the local society. It acts under the auspices of the Finnish Mining Industry (FinnMin). To build responsible and sustainable mining practices, tools suitable for Finland have been developed and tailored. Synergies between different industries are emphasized to prevent conflicts of interest. Furthermore, information and experience are shared, and cooperation encouraged. The Network tries to influence its members such as by asking companies that are implementing a mining liability and consequently carrying out self-assessments to share the results with the Network.

The mining liability system and its training material serve as the operating principles of the industry, and a guideline for when the system's assessment tools are applied. The self-assessments are verified every three years, but verification from an external expert is needed before the results are submitted to the Network. The Network has expressed its ambition for Finland to become a pioneer country in sustainable mining. In this regard, the industry has to take into account the natural environment, cultural and social aspects, and other livelihoods (Kaivosvastuu, n.d.a, b, d).

The operating principles of the mining liability systems are the following (Kaivosvastuu, n.d.c):

- The companies' operations are **transparent**.
- The companies follow **best practices** in the mining industry and are committed to **continuous improvement**.
- The companies **minimize the negative impacts of their operations on** nearby communities, the environment, and biodiversity.
- Companies **respect the surrounding community** and take into account the culture and lifestyles of the community.
- The companies promote projects that safeguard **the viability of the** mining area **after** the mining operations cease.
- Companies actively maintain a **dialogue with stakeholders**.
- Companies do not endanger **the health and safety** of their employees and nearby communities.
- The companies promote **the sustainable use of mineral resources** and participate in development projects aimed at the safe and responsible production and recycling of metals and minerals.
- The companies **are closing the mines responsibly** and supporting the development of closure operations.

Collectively, the Network has published various documents. These include a collective 2014 CSR report on 19 companies operating in Finland within the industry, a toolbox for the local action of companies emerging in the field of ore exploration or mining in specific Finnish regions, a summary report of activities of the Network during its first years of operation, and a sustainability standard for mining. Since the publication of the initial CSR report of the Network, it has collected reports from its members annually and publishes them on its website. The purpose of the publication is transparency about the industry operations and its impact on the environment and society; the content of the reports is both general and company-specific. The toolbox provided by the Network is twofold, one for mining and another one for ore finding (Kaivosvastuu, n.d.b).

The *Towards Sustainable Mining (TSM)* standard builds on a similar standard from Canada but has been adapted in accordance with Finnish legislation. It complements two protocols on water management and mine closure so that it now covers the whole lifecycle of mining operations (Kaivosvastuu, n.d.b). Additionally, the Finnish TSM includes protocols in the area of community outreach, biodiversity conservation, tailings management, energy use and greenhouse gas emissions, health and safety, and crisis management (Kaivosvastuu, n.d.b). These protocols are available on the website of the Sustainable Mining Network and include assessment methods, guidelines, performance indicators, and self-assessment checklists (Kaivosvastuu, n.d.b).

This case demonstrates that the collective actions of an industry or a network can influence sustainable best practices and at the same time save resources, for the benefit of the whole. A collective CSR industry report may also result in comparability of findings within the industry and improvements in the content of reports and indicators, but this means that these reports have to be published regularly.

Iceland

Iceland is in northern Europe, located between mainland Europe and North America. The landmass of the island is 103,000 sq km, and the population is around 365,000 inhabitants (Central Intelligence Agency, n.d.d; Statistics Iceland, 2020). The Arctic Circle passes through the island of Grímsey, Iceland's northernmost community located 40 kilometres off the north coast of Iceland. Iceland is the only Arctic State that does not have an Indigenous population (Arctic Council, n.d.d).

The main sectors of the Icelandic economy are services, mainly tourism, manufacturing industries, construction and utilities, and fisheries (Promote Iceland, n.d). The key industries rely on sustainable utilization of natural energy and marine resources, specifically energy-intensive industries, fisheries, and tourism (Arctic Council, n.d.d; Promote Iceland, n.d).

Tourism, energy, and fisheries characterize the Icelandic economy, and per capita, the nation is the world's largest producer of electricity (Glomsrød et al., 2017). Extensive use of renewable energy is one of the characteristics of the Icelandic economy. Renewables, mainly hydro and geothermal energy, comprised 88.5% of energy consumption in 2017, much higher than the average primary energy supply in the OECD countries (OECD, 2019). Greenhouse gas emissions are, nevertheless, substantial as most of the energy is used by energy-intensive industries such as aluminium smelting. These emissions are incorporated in the European Union's (EU) emission trading scheme, although Iceland is not an EU member state. Emissions also stem from agriculture and sheep raising, and consequent soil erosion (OECD, 2019). Of value added by industries in Iceland, the public administration has the highest share of value added, followed by service activities, real estate activities, wholesale and retail trade, manufacturing, and financial and insurance services (Glomsrød et al., 2017).

Carbfix: A case study from Iceland

Carbfix started as a research project in 2006, receiving funds from the EU's FP7 and Horizon 2020 research and innovation programmes. The project was formalized in 2007 by its founding partners, Reykjavík Energy, the University of Iceland, CNRS in Toulouse, and the Earth Institute at Columbia University. The preparation stage of the company lasted from 2006 until 2011. The pilot phase started in 2012, and lasted until 2016, overlapping with the industrial-scale operation starting in 2014. In 2020, Carbfix became a subsidiary of Reykjavik Energy (OR), an Icelandic energy and utility company (Carbfix, n.d.c). Carbfix Ltd. is one of the world's first companies dedicated to the geological storage of CO_2.

The key idea of Carbfix is to capture CO_2 emissions and other soluble gases, and permanently store them in the subsurface where it takes them less than two years to transform into minerals. Through this process, Carbfix has demonstrated how CO_2 emissions can be reduced in a cost-effective, secure, and environmentally friendly way (Carbfix, n.d.d). The technology has also been used to capture direct CO_2 emissions from the air in collaboration with the Swiss company Climeworks. The activities of Carbfix were separated from the operations of Reykjavik Energy (OR) for several reasons, such as to protect Carbfix intellectual property rights and to strengthen its core business, which is to give stakeholders in Iceland and abroad the chance of reducing their carbon footprint by using the Carbfix solution (Carbfix, n.d.b).

So far, the cost of the carbon capture and storage chain using the CarbFix method is ~$25/ton, compared to the $26.50/ton carbon quota price on the EU carbon trading scheme market. It is anticipated that further development of the project will reduce costs for capturing emissions from various power and industrial processes (Carbfix, n.d.b). The novelty of the technology recently earned it the Ruggero Bertani European Geothermal Innovation Award 2020. On the website of the European Geothermal Energy Council (EGEC), it is stated that the Carbfix process "overturns the common view that mineral storage in Carbon Capture & Storage projects takes hundreds to thousands of years. Carbfix has global scaling potential in and beyond the geothermal industry for decarbonising the economy". The project is also among 20 world-class projects nominated as the finalists for the 2020 Keeling Curve Prize, which locates and funds the world's most promising climate solutions; it is nominated in the carbon capture and utilization category (Global Warming Mitigation Project, 2020).

The company has plans for lowering costs and adding value to the method by capturing simultaneously other greenhouse gases and injecting them into the ground, expanding direct air capture application of the solution, moving offshore with the solution by using sub-marine basalts and seawater to capture and store the gases, and exporting the solution to other geological sites, starting within Europe (Carbfix, n.d.a).

This case demonstrates that an innovative idea that is supported and nurtured can result in a solution of global importance.

Norway

Norway is in northern Europe, located west of Sweden. It borders the North Sea and the North Atlantic Ocean. The landmass is a total of 323,802 sq km and the total population is around 5.5 million inhabitants (Central Intelligence Agency, n.d.e). Nearly half of the Norwegian landmass is an Arctic territory. These are the combined County of Troms and Finnmark, the Nordland County, and the island of Jan Mayen and the Svalbard archipelago (Arctic Council, n.d.e). The Arctic Indigenous Peoples are Saami. The total population in the Norwegian Arctic is 490,000 inhabitants; of around 80,000 Saami Peoples living in Norway, Sweden, Finland, and Russia, around half live in Norway (Arctic Council, n.d.e; NordNorsk Reiseliv AS, n.d.).

The cornerstone of the economy in the northern part of Norway is fisheries, marine resources, and livestock husbandry. The economy is, however, becoming more diverse. Although fisheries and aquaculture are the most critical sectors with regard to exports, tourism, sustainable energy (hydro and wind power), power from liquefied gas, minerals, and transit are also becoming important (Arctic Council, n.d.e). In terms of value added by industries, most of the value added is associated with health care and social work, public administration and defence, education, real estate activities, construction, and wholesale and retail trade (Glomsrød et al., 2017).

The Association of Arctic Expedition Cruise Operators: Case study from Norway

The purpose of the Association of Arctic Expedition Cruise Operators, among others, is to "ensure that expedition cruises and tourism in the Arctic is carried out with the utmost consideration for the vulnerable, natural environment, local cultures and cultural remains, as well as the challenging safety hazards at sea and on land" (AECO, n.d.b). The Association was founded in 2003, and its office is in Tromsø, Norway. It is an international association for expedition cruise operators (AECO, n.d.a). The geographical range of the Association is "the Arctic area north of 60 degrees north latitude. The core areas are Svalbard, Jan Mayen, Greenland, Arctic Canada, the Russian Arctic National Park, and Iceland" (AECO, n.d.a).

Members of the Association are obliged "to operate in accordance with national and international laws and regulations and have, in addition, agreed to follow an extensive set of guidelines to ensure operations are in accordance with our objectives, including a number of site-specific guidelines and guidelines for visitors to the Arctic" (AECO, n.d.c). The AECO members operate in accordance with guidelines relating to the natural environment,

Table 4.2 Overview of AECO guidelines for its members

Environmental guidelines	*Social and cultural guidelines*	*Operations of AECO members*
Clean Seas guidelines	Visitor guidelines	Operational guidelines
Vegetation guidelines	Community guidelines	Yacht guidelines
Wildlife guidelines	Cultural remains guidelines	
Biosecurity guidelines	Site guidelines	

Source: AECO (n.d.c)

social and cultural aspects of Arctic communities and sites, and operations of AECO members (see Table 4.2). These guidelines are the backbone of AECO. Additionally, posters on wildlife, birds, communities, and responsible tourism displayed on vessels sailing in the Arctic are used to inform guests about responsible Arctic tourism. Expedition leaders and guides can test their knowledge of relevant regulations and guidelines by taking online exams provided by AECO (AECO, n.d.c).

As an example, the visitor guidelines are published in three languages, namely English, German, and Norwegian, but videos of the same guidelines available for AECO members have subtitles in 14 languages. In the visitor guidelines, the responsibilities of visitors and AECO are both emphasized. Visitors should follow eight principles (AECO, n.d.e):

1. Leave no lasting signs of your visit.
2. Do not pick flowers.
3. Do not take anything with you.
4. Do not disturb animals or birds.
5. Leave cultural remains alone.
6. Take the polar bear danger seriously.
7. Respect local culture and local people.
8. Be safe.

The importance of sensitive cultural and social interaction is particularly stressed, as well as safety precautions related to polar bears and Arctic dogs. In the case of cultural and social interaction, visitors are expected to respect local cultures of the Inuit population, which is regarded as a part of sustainable tourism in the Arctic (AECO, n.d.e).

Additionally, AECO offers on its website an overview of relevant resources and tools for Arctic cruise operators. These include relevant databases, meteorological and ice information, maps, forms, links such as related to search and rescue, acts, regulations and permits, publications, and links to useful local and/or national information including relevant national or local authorities, institutions, and associations (AECO, n.d.d).

Given the impacts of the COVID-19 situation on Arctic cruise ship and expedition operators, AECO provides its members with regional updates on the situation. These include information about the population in a country or a region, confirmed cases in these locations, recent updates from governments and authorities such as regarding travel restrictions, travel advice and information from authorities, links to official health authorities' websites on the coronavirus, and prospects for visits to the relevant locations (AECO, n.d.a).

This case demonstrates the importance of an association in developing and utilizing voluntarily non-regulatory guidelines developed for its members to follow in order to operate responsibly and respectfully in a sensitive natural, social, and cultural environment. Guidelines provided serve as best-practice examples.

Sweden

Sweden is in northern Europe, between Finland and Norway. It borders the Baltic Sea, the Gulf of Bothnia, Kattegat, and Skagerrak. The total landmass is 450,295 sq km, and the total population is around 10.2 million inhabitants (Central Intelligence Agency, n.d.f). The Norrbotten County and the Västerbotten County are Arctic territories, covering around one-third of the Swedish landmass. Approximately 520,000 persons live in the Swedish Arctic. The Arctic Indigenous Peoples are Saami, with around 20–50,000 living in the Swedish Arctic (Arctic Council, n.d.f).

The main Swedish export commodities are industry machinery and road vehicles (15% in each category), electronics and telecommunication (10%), petroleum products (7%), food, beverages, and tobacco, paper and paper products, and pharmaceutical products (6% in each category), and non-specified (35%) (OECD, 2019). Arctic Sweden is characterized by a strong emphasis on the manufacturing industry, more specifically basic metals and wood and paper. Mining is also of great importance, as well as utilities (i.e. electricity production, gas, and water supply), tourism, and other private services. In terms of value added by industries, non-specific service activities have the highest value, followed by health care and social work, manufacturing, mining and quarrying, and utilities (Glomsrød et al., 2017).

Tentipi: A case study from Sweden

Tentipi is a company established in 1989. The founder of the company, Bengt Grahn, was staying in the village of Moskosel in Swedish Lapland, near the Arctic Circle, when he got the idea of making Nordic tipis. At the beginning of the process, the entrepreneur did everything himself, from purchasing the material and sewing the tents, to marketing and selling (Tentipi, 2020). The first office and the manufacturing site of the company were in the village of Moskosel, where many of the tents are still made today (Tentipi, 2019, 18 April). The idea behind the company and its products is founded on Sami heritage where people socialize around a fire. This design is inspired by the kåta, which are the traditional homes of semi-nomadic Sami people. The cone-shaped frame of tipis are made of wooden poles, overlaid by animal skins, and open at the top so the smoke can seam out. The traditional methods are combined with new innovative designs, such as particular construction techniques and the latest materials. The camping tents are thought of as homes, not shelters, and can be used all year round (Tentipi, n.d.).

Initially, the company produced small tents, but then received orders for giant Nordic tipis. After initial difficulties in designing and producing these giant tipis, the tents were redesigned based on innovative ideas resulting in several patents and design rights. In 1993, the company was approached by a Sami cultural organization, who asked it to create an exhibition space (25×50m) to promote Sami culture and products (Tentipi, 2020). Other innovative solutions are to open the sides of the tents and expansion wedges. At first, the giant tents were named “Giant Hats” and were later renamed “Stratus”. Another innovative idea was to link the Stratus tents together and linking technology was developed, allowing the tents to link together in various forms covering a large area, so they could host up to 2,000 people for dinner. The demand for these tents started to increase and they are now sold in every corner of the world (Tentipi, 2019, 16 January), where they are used for various events, such as weddings and music festivals (Tentipi, 2020).

Development is still ongoing and three aims are fundamental:

> 1) to introduce new products to use alongside the Stratus, extending its relevance to a wider and wider range of events; 2) to invent new ways to build, dismantle and link the Stratus more quickly without compromising safety; 3) to keep ahead of developments in standards relating to fire retardancy, structural stability and environmental impact
>
> (Tentipi, 2020)

This case demonstrates that an innovative idea, inspired by Indigenous culture, can result in a successful global business.

The Russian Federation

The Russian Federation is in north Asia, with borders extending from Europe to the North Pacific Ocean. The total landmass is 17,098,242 sq km, and the total population is approximately 142 million inhabitants (Central Intelligence Agency, n.d.g). The Arctic territory of the Russian Federation includes Murmansk, Nenets, Yamal-Nenets, Chukotka Autonomous Okrugs, and northern municipalities of Arkhangelsk, Komi Republic, Krasnoyarsk Territory, and Republic of Sakha (Yakutia) (Arctic Council, n.d.g). The Russian Arctic population is around 2.5 million inhabitants; 11 groups of Indigenous Peoples live around or above the Arctic Circle, comprising approximately 270,000 inhabitants in the Arctic (Arctic Council, n.d.g, h).

The largest Indigenous Peoples groups are the Dolgan, Nganasan, Nenets, Saami, Khanty, Chukchi, Evenk, Even, Enets, Eskimo (Yupik), and Yukagir, which rely on fishing, hunting, and reindeer herding for their livelihood (Arctic Council, n.d.g).

In Arctic Russia, petroleum and other mining industries are of critical importance, with a high dependency on primary production, predominantly oil, gas, and other minerals. In other cases, value added by industries comes mainly from construction, transportation and storage, wholesale and retail trade, and real estate activities (Glomsrød et al., 2017). Arctic economic development in Russian territories requires investments in extraction, transportation, and governance infrastructure (GeoHistory Today, 2018), and the development of the northern shipping industry (Plotnikov et al., 2018). Military activities are also highlighted as important aspects of Russia's Arctic development (GeoHistory Today, 2018). The highest value added by industry in Arctic Russia is by far from petroleum and other mining activities, followed by construction, transportation and storage, wholesale and retail trade, and real estate activities (Glomsrød et al., 2017).

Novatek: A case study from Russia

CSRHub ranks the Russian natural gas company PAO Novatek, or Novatek for short, in Yamal among the leaders with regard to CSR and environment, social and governance (ESG) metrics. The scale used provides absolute numbers from 0 (low) to 100 (high). Among 19,495 companies ranked by the CSRHub, Novatek's score is 75. The categories used for ranking companies are community, employees, environment, and governance, each with three subcategories; see Table 4.3 (CSRHub, 2020).

Novatek was founded in 1994 and it explores, processes, and produces and processes natural gas and liquid hydrocarbons (CSRHub, 2020). It is the largest independent natural gas producer in Russia. The company's strategy has four main categories: conservative financial policies, corporate governance, efficient investment decisions, and sustainable development. It's

Table 4.3 CSRHub ranking categories

Community	*Employees*	*Environment*	*Governance*
Community Development & Philanthropy	Compensation & Benefits	Energy & Climate Change	Board
Product	Diversity & Labor Rights	Environment Policy & Reporting	Leadership Ethics
Human Rights & Supply Chain	Training, Health & Safety	Resource Management	Transparency & Reporting

Source: CSRHub (2020)

resource-based growth is based on innovation and technology, achieved by maintaining low-cost structure, an increase of hydrocarbon production, optimization of market channels, and by building a low-cost scalable liquefied natural gas (LNG) platform (Novatek, 2020a).

In 2019, Novatek received the *Change Management. Visionaries* award. This is "Russia's first-ever professional award in sustainability impact assessment. Its main goal is to promote the implementation of impact-oriented approaches in social and environmental decision-making". The company's Sustainability Report was recognized as the winner in the category "Disclosure of Information on Corporate Governance" (Novatek, 2019). On this occasion, Novatek issued the following statement (Novatek, 2019):

> "Sustainable development is integral to our corporate strategy and a cornerstone of a successful, responsible business conduct," noted NOVATEK's Deputy Chairman of the Management Board Mark Gyetvay. "Each year we strive to improve the quality and quantity of information disclosed in our sustainability reports in line with the best international practices. We are one of the major players in the global LNG market, and we are fully committed to maintaining the highest standards in corporate governance and transparency".

More specifically, sustainable development within the company is subcategorized on its website into health and safety, social responsibility, payments to governments, archives, environmental protection, and integrated management system (Novatek, 2020b). Information disclosed in the awarded Sustainability Report is in line with the Global Reporting Initiative (GRI) standards. It highlights Novatek's ESG principles and practices as serving as a channel for stakeholder engagement. The report itself complies with GRI reporting principles, namely balance, comparability, accuracy, timeliness, clarity, and reliability (Novatek, 2020c). Novatek incorporates into its internal regulations the universal principles of the International Labor Organization's Declaration on Fundamental Principles and Rights at Work, the UN Sustainable Development Goals (SDGs) of the 2030 Sustainability Agenda, and best practices of disclosing its payments to governments as presented in the Extractive Industries Transparency Initiative (EITI) global standard (Novatek, 2020c).

To name a few examples, in regard to environmental responsibility the company focuses on mitigating its environmental impact, sustainable use of natural resources and land, exercising reasonable care in implementing projects, biodiversity protection, investment in the low-carbon economy, and compliance with global environmental standards. In the case of local development, the focus is on improving living standards for local communities, supporting local communities, including Indigenous minorities of the North, social partnership with regional authorities, and educational and cultural programmes. Regarding ethical business practices, Novatek stresses its emphasis on a zero-corruption policy, effective and transparent corporate governance framework, respect of human rights and freedoms, and good business reputation (Novatek, 2020c).

Most of the 17 SDGs are of relevance to Novatek's operations. The goals not addressed specifically in the 2018 Sustainability Report are Goal 2, zero hunger, and Goal 5, gender equality. The company also recognizes major risk factors that may affect the business, including sustainability risks, and mitigation of risk factors such as through insurance. Operational risk factors include sustainability-related risks, such as a risk of early termination, suspension, or restriction of the right to use subsurface mineral resources, and environmental, social, and ethical risks. Financial risk factors include credit, liquidity, and inflation risks, and legal risk factors include risk of law changes and sanctions and litigation risk (Novatek, 2020c).

This case highlights the importance of transparency and recognition of not just opportunities, but more importantly risk factors, for companies operating in the non-renewable energy sectors, as these risk factors may influence their operation conditions and thus investors, authorities and legislation, the natural environment, and the social and cultural circumstances where operations take place.

The United States

The United States (US) is in North America, between Canada and Mexico, bordering the North Atlantic Ocean and the North Pacific Ocean. The total landmass is 9,833,517 sq km and the population is around 333 million inhabitants (Central Intelligence Agency, n.d.h). The US Arctic territory is "north and west of the boundary formed by the Porcupine, Yukon, and Kuskokwim Rivers; all contiguous seas, including the Arctic Ocean and the Beaufort, Bering and Chukchi Seas; and the Aleutian chain" (Arctic Council, n.d.i). The US Arctic population is approximately 50,000 inhabitants (Arctic Council, n.d.i), but the figures on the number of Arctic Indigenous Peoples in the US are somewhat unclear. Available numbers state that the Indigenous populations are between 2.5 and 6 million people, based on self-identification and different sampling methods. Of this population, 23% live in Alaska Native villages or American Indian areas (IWGIA, 2019, 2020), but no further categorization is made. The US Arctic Indigenous population consists of the Aleut, Alutiiq, Yup'ik, Iñupiaq (Northwest Alaskan Inuit), Athabaskan, Tlingit, and Haida (Arctic Council, n.d.i).

In Alaska, the main industries, in terms of value added, are mining and quarrying (i.e. oil and gas extraction, and other mining and quarrying activities), meaning that petroleum is the backbone of the economy. Other important economic activities are public administration and defence, transportation by pipeline and storage, and unspecified service activities. The seafood industry in Alaska and tourism are also worth mentioning (Glomsrød et al., 2017).

The Red Dog mine: A case study from the United States

Teck Resources Limited operates the Red Dog Mine, one of the largest zinc mines in the world (Teck, 2017). Teck Resources Limited collaborates with the for-profit Northwest Alaska Native Association (NANA) who owns the land. NANA serves more than 14,500 local stakeholders and traditional communities of the NANA region in northwest Alaska (NANA, n.d.). The purpose of the collaboration is "to operate responsibly by creating a safe and sustainable environment" both for the people and the region (Teck, 2017). Some of the key milestones in the history of the development are (North of 60 Mining News, 2015; Teck, 2017):

- 1982: Agreement made between NANA and Cominco (now Teck) on the mining operation and benefit sharing.
- 1986: Initial development began with the installation of a shallow water dock and small staging area at the port site.
- 1987: Major construction on the road and mine site started.
- 1989: Mill largely built in modules in the Philippines, transported by barge to the coast and module movers to site.
- 1989: Construction completed.

More than half of the employees at the Red Dog mine are NANA shareholders, both regular employees (52%) and seasonal and temporary (93%). Teck offers educational opportunities for employees, both during elementary and high school, namely career awareness and job shadowing opportunities, summer students and intern programme, and scholarships for university-level students (Teck, 2017). The PIC abbreviation, which stands for *prevention*, continuous *improvements*, and *compliance*, provides an example of other types of commitment concerning the natural environment (pollution and waste), health and safety risks, commitments made to the community, and compliance with laws, regulations, and permits (Teck, 2017). In order to respect and protect the subsistence lifestyle a subsistence committee was formed. It is involved in operation, construction, and exploration activities; mine water discharge and drinking water issues; issues related to fish, caribou, and marine mammals; air quality and dust prevention; shipping season; major projects; and planning of closure (Teck, 2017). Furthermore, significant community engagement takes place, as well as community advancement, such as through a Village Wild Seed Collection Program. The company makes long-term commitments in its sustainability strategy, as some of its goals stretch through to 2050 in eight strategic themes: 1) communities and Indigenous Peoples, 2) biodiversity and reclamation, 3) water, 4) climate change, 5) responsible production, 6) our people, 7), tailings management, and 8) health and safety (Teck, n.d.a).

Teck publishes annually several types of sustainability reports. In the year 2019, the company published the 2019 Sustainability Report, Sustainability Performance Data, 2019 Sustainability Accounting Standards Board (SASB) Index, UN Global Compact (UNGC) Report, and more (Teck, n.d.b). The material topics reported relate to some of the strategic themes: health and safety, water stewardship, relationship with Indigenous Peoples, relationship with communities, tailings management, climate change, energy use, diversity and employee relations, air quality, supply-chain management, responsible production and waste management, biodiversity and reclamation, environmental management, human rights, and business ethics (Teck, n.d.b). The company is very open about its reporting practices, and everyone can assess their sustainability performance data by downloading an Excel spreadsheet from their website (Teck, n.d.b). Also, it updates its policies as needed, such as in the case of COVID-19, where the operation's travel policies were updated in order to safeguard the health of employees and communities (Teck, 2020a, b).

The UN Global Compact Reports of Teck are, and have been since 2014, marked as advanced according to UNGC. Three categories (GC Learners, GC Active, and GC Advanced) are used for the Communication on Progress (CoP) reports, based on three minimum requirements:

1. A statement by the Chief Executive expressing continued support for the UN Global Compact and renewing the participants' ongoing commitment to the initiative.
2. A description of practical actions the company has taken or plans to take to implement the Ten Principles in each of the four areas (human rights, labour, environment, anti-corruption).
3. A measurement of outcomes.

Based on self-assessment, the Global Compact means that reports meet the above requirements, and additionally include information on the implementation of advanced criteria and best practices. This suggests Teck's ambition regarding transparency and sustainability reporting (UN Global Compact, 2020; n.d.).

This case demonstrates is the importance of sharing the benefits between the Indigenous Peoples and the company involved in a mining project. Transparency about the operation, both in terms of positive and negative aspects of the operation, is of critical importance if trust is to be established and maintained.

Questions for discussion

1. How can you identify if CSR activities of companies are genuine or not?
2. How can local communities benefit from large-scale business operations?

3. What are the risks to local communities if their interests are too intertwined with the interests of the main company operating in the area?
4. What are the risks to the natural environment if the interests of local communities are too intertwined with the interests of the main company operating in the area?
5. How important is CSR for companies in controversial industries?

Further resources

Many other cases may be interesting, such as in relation to the management of renewable biological resources. The Nordic Council of Ministers (2017) has published a report including "25 Nordic cases on the 'new' Nordic bioeconomy". This topic is explored in more detail in Chapter 7.

Chapter summary

The key focus of this chapter was the Arctic nations and their respective industries. For each nation, a selected case was discussed (see Table 4.1), with the cases ranging from fine art and culture to mining and hydrocarbons, from small newly founded companies to large industry groups and networks, and from cooperatives to government-owned and limited liability companies, thus moving the discussion from theory to practice.

Each case brought forth some specific learnings. The West Baffin Eskimo Co-operative case (Canada, Nunavut) illustrated how nurturing local culture can create an income stream for local artists and their families through an international reach of exquisite Inuit artwork known for its originality and quality. The Royal Greenland case (Denmark/Greenland) demonstrated how local government can run a successful business, taking into account both the latest modern technology and traditional fishing methods, thus utilizing marine resources responsibly and sustainably while maximizing economic value for the benefit of society. The Sustainable Mining Network case (Finland) demonstrated that the collective actions of an industry or a network can influence sustainable best practices and at the same time secure financial resources, for the benefit of the whole. The Carbfix case (Iceland) illustrated that an innovative idea from one of the key industries, in this case, renewable energy resources, that is supported and nurtured can result in a solution to the climate crisis that is of global importance. The case of the Association of Arctic Expedition Cruise Operators (AECO) (Norway, Tromsø) demonstrated the importance of an association in developing and utilizing voluntarily non-regulatory guidelines for its members to follow in order to operate responsibly and respectfully in a sensitive natural, social, and cultural environment. Guidelines provided serve as best-practice examples. The Tentipi case (Finland) illustrated that an innovative idea, inspired by Indigenous culture, can result in a successful global business. The Novatek (Russia, Yamal) case highlighted the importance of transparency and recognition of not just opportunities, but more importantly risk factors, for companies operating in the non-renewable energy sectors, as these risk factors may impact their operation conditions and consequently investors, authorities and legislators, the natural environment, and the social and cultural circumstances where operations take place. The Red Dog Mine (United States, Alaska) case demonstrated the importance of sharing the benefits between the Indigenous Peoples and the company involved in a mining project. Transparency about the operation, both in terms of positive and negative aspects of the operation, is of critical importance if trust is to be established and maintained.

References

Alaska Community Action on Toxics. (2020). Mining in Alaska. Accessed on 13 July 2020 from www.akaction.org/tackling_toxics/alaska/mining_in_alaska/

Arctic Co-operatives Limited. (n.d.). Arctic Co-operatives Limited. Accessed on 10 July 2020 from https://arctic-coop.com/

Arctic Council. (n.d.a). Canada. Accessed on 15 May 2020 from https://arctic-council.org/en/about/states/canada/

Arctic Council. (n.d.b). Denmark. Accessed on 22 April 2020 from https://arctic-council.org/en/about/states/denmark/

Arctic Council. (n.d.c). Finland. Accessed on 15 May 2020 from https://arctic-council.org/en/about/states/finland/

Arctic Council. (n.d.d). Iceland. Accessed on 15 May 2020 from https://arctic-council.org/en/about/states/iceland/

Arctic Council. (n.d.e). Norway. Accessed on 18 May 2020 from https://arctic-council.org/en/about/states/norway/

Arctic Council. (n.d.f). Sweden. Accessed on 18 May 2020 from https://arctic-council.org/en/about/states/sweden/

Arctic Council. (n.d.g). The Russian Federation. Accessed on 20 May 2020 from https://arctic-council.org/en/about/states/russian-federation/

Arctic Council. (n.d.h). Russian Association of Indigenous Peoples of the North. Accessed on 20 May 2020 https://arctic-council.org/en/about/permanent-participants/raipon/

Arctic Council. (n.d.i). The United States. Accessed on 26 May 2020 from https://arctic-council.org/en/about/states/the-united-states/

Arctic Economic Council (AEC). (n.d.). About us. Accessed on 8 June 2020 from https://arcticeconomiccouncil.com/about-us/

Arctic Institute. (n.d.). Finland. Accessed on 15 May 2020 from www.thearcticinstitute.org/countries/finland/

Association of Arctic Expedition Cruise Operators (AECO). (n.d.a). AECO. Accessed on 10 July 2020 from www.aeco.no/

Association of Arctic Expedition Cruise Operators (AECO). (n.d.b). About us. Accessed on 10 July 2020 from www.aeco.no/about-aeco/

Association of Arctic Expedition Cruise Operators (AECO). (n.d.c). Guidelines. Accessed on 10 July 2020 from www.aeco.no/about-aeco/

Association of Arctic Expedition Cruise Operators (AECO). (n.d.d). Resources & tools. Accessed on 10 July 2020 from www.aeco.no/resources-and-tools/

Association of Arctic Expedition Cruise Operators (AECO). (n.d.e). Visitor guidelines. Accessed on 10 July 2020 from www.aeco.no/guidelines/visitor-guidelines/

Arruda, G.M. (2019). *Sustainable Energy Education in the Arctic: The Role of Higher Education* (1st ed.). Abingdon, UK: Routledge.

Carbfix. (n.d.a). *Carbfix – Fact Sheet*. Reykjavík, Iceland: Carbfix.

Carbfix. (n.d.b). Carbfix as a subsidiary. Accessed on 9 June 2020 from www.carbfix.com/carbfix-a-subsidiary

Carbfix. (n.d.c). Our story. Accessed on 9 June 2020 from www.carbfix.com/our-story

Carbfix. (n.d.d). What is Carbfix? Accessed on 9 June 2020 from www.carbfix.com/

Central Intelligence Agency. (n.d.a). The world factbook. Canada. Accessed on 15 May 2020 from www.cia.gov/library/publications/the-world-factbook/geos/ca.html

Central Intelligence Agency. (n.d.b). The world factbook. Greenland. Accessed on 22 April 2020 from www.cia.gov/library/publications/the-world-factbook/geos/gl.html

Central Intelligence Agency. (n.d.c). The world factbook. Finland. Accessed on 15 May 2020 from www.cia.gov/library/publications/the-world-factbook/geos/fi.html

Central Intelligence Agency. (n.d.d). The world factbook. Iceland. Accessed on 15 May 2020 from www.cia.gov/library/publications/the-world-factbook/geos/ic.html

Central Intelligence Agency. (n.d.e). The world factbook. Norway. Accessed on 18 May 2020 from www.cia.gov/library/publications/the-world-factbook/geos/no.html

Central Intelligence Agency. (n.d.f). The world factbook. Sweden. Accessed on 18 May 2020 from www.cia.gov/library/publications/the-world-factbook/geos/sw.html

Central Intelligence Agency. (n.d.g). The world factbook. The Russian Federation. Accessed on 20 May 2020 from www.cia.gov/library/publications/the-world-factbook/geos/rs.html

Central Intelligence Agency. (n.d.h). The world factbook. United States. Accessed on 26 May 2020 from www.cia.gov/library/publications/the-world-factbook/geos/us.html

CSRHub. (2020). Corporate Social Responsibility (CSR) & Environment, Social, Governance (ESG) Metrics – Novatek CSR/ESG Ranking. Accessed on 10 July 2020 from www.csrhub.com/CSR_and_sustainability_information/Novatek

Dorset Fine Arts (DFA) (n.d.). Dorset Fine Arts. Accessed on 10 July 2020 from www.dorsetfinearts.com/

European Geothermal Energy Council (EGEC). (2020). Carbfix and Eavor, geothermal innovators of the year. Accessed on 9 June 2020 from www.egec.org/carbfix-and-eavor-geothermal-innovators-of-the-year/

GeoHistory Today. (2018). Russia's Arctic Development: Problems and Priorities. Accessed on 8 June 2020 from https://geohistory.today/russia-arctic-development-power/

Global Warming Mitigation Project. (2020). 2020 Finalists. Accessed on 9 June 2020 from www.globalwarmingmitigationproject.org/2020-finalists

Glomsrød, S., Duhaime, G., and Aslaksen, I. (eds.). (2017). *The Economy of the North 2015*. Oslo, Norway: Statistisk sentralbyrå.

Glomsrød, S., Mäenpää, I., Lindholt, L., McDonald, H., Wei, T., and Goldsmith, S. (2017). Arctic economies within the Arctic nations. In S. Glomsrød, G. Duhaime, and I. Aslaksen (Eds.), *The Economy of the North 2015* (pp. 37–78). Oslo, Norway: Statistisk sentralbyrå.

Government of Canada. (2016). *Canadian Arctic Capabilities*. Ottawa, Canada: Author.

Government of Greenland. (n.d.). Economy and industry in Greenland. Accessed on 15 May 2020 from https://naalakkersuisut.gl/en/About-government-of-greenland/About-Greenland/Economy-and-Industry-in-Greenland

GRIDA. (2005). Economy of the Arctic, by sector. Accessed on 27 May 2020 from www.grida.no/resources/7001

International Work Group for Indigenous Affairs (IWGIA). (2019). Indigenous world 2019: The United States of America. Accessed on 26 May 2020 from www.iwgia.org/en/usa/3375-iw2019-usa.html

International Work Group for Indigenous Affairs (IWGIA). (2020). Indigenous peoples in the United States. Accessed on 26 May 2020 from www.iwgia.org/en/usa.html

Kaivosvastuu. (n.d.a). Authentication. Accessed on 9 July 2020 from www.kaivosvastuu.fi/todentaminen/

Kaivosvastuu. (n.d.b). Finnish sustainability standard for mining translated into English. Accessed on 9 July 2020 from www.kaivosvastuu.fi/finnish-sustainability-standard-for-mining-translated-into-english/

Kaivosvastuu. (n.d.c). Training material: Mining responsibility system. Accessed on 9 July 2020 from www.kaivosvastuu.fi/handbook-3/

Kaivosvastuu (n.d.d). What is the Sustainable Mining Network? Accessed on 9 July 2020 from www.kaivosvastuu.fi/

Mäenpää, I., Glomsrød, S., and Wei, T. (2017). Comparative analysis of Arctic economies at macro level. In S. Glomsrød, G. Duhaime, and I. Aslaksen (Eds.), *The Economy of the North 2015* (pp. 27–36). Oslo, Norway: Statistisk sentralbyrå.

NANA. (n.d.). About us – Alaska Native Corporation. Accessed on 13 July 2020 from https://nana.com/about-us/

Nordic Council of Ministers (2017). Nordic bioeconomy – 25 cases for sustainable change. Accessed on 13 July 2020 from www.norden.org/en/news/25-nordic-bioeconomy-cases-sustainable-change

NordNorsk Reiseliv AS. (n.d.). The Sami: Indigenous people of the North. Accessed on 18 May 2020 from https://nordnorge.com/en/tema/the-sami-are-the-indigenous-people-of-the-north/

Nordregio. (n.d.). Indigenous population in the Arctic. Accessed on 26 May 2020 from https://nordregio.org/maps/indigenous-population-in-the-arctic/

Novatek. (2020a). Company – Business strategy. Accessed on 10 July 2020 from www.novatek.ru/en/about/strategy/

Novatek. (2020b). Sustainable development. Accessed on 10 July 2020 from www.novatek.ru/en/development/

Novatek. (2020c). Sustainable development – Sustainability Report 2018. Accessed on 10 July 2020 from www.novatek.ru/en/development/

Novatek. (2019). Novatek wins the Sustainability Impact Assessment Award. Accessed on 10 July 2020 from www.novatek.ru/en/press/releases/index.php?id_4=3151

North of 60 Mining News. (2015). 25 years of sustainability at Red Dog. Accessed on 13 July 2020 from www.miningnewsnorth.com/story/2015/07/05/news-nuggets/25-years-of-sustainability-at-red-dog/3247.html

Organisation for Economic Co-operation and Development (OECD). (2005). Country note on national fisheries management system – Greenland. Accessed on 26 May 2020 from www.oecd.org/countries/greenland/34431581.pdf

Organisation for Economic Co-operation and Development (OECD). (September 2019). OECD Economic Survey – Iceland. Accessed on 26 May 2020 from www.oecd-ilibrary.org/economics/oecd-economic-surveys-iceland-2019_c362e536-en

Plotnikov, V., Kutepova, M., and Sushko, O. (2018). The economy of the Russian Arctic: State and specifics of development. *Advances in Economics, Business and Management Research (AEBMR)*, 39, 706–710.

Promote Iceland. (n.d.). Economy & infrastructure. Accessed on 15 May 2020 from www.iceland.is/the-big-picture/economy-infrastructure

Rautajoki, T. and Lakkapää, V. (2019). *Arctic Business Forum Yearbook 2019*. Rovaniemi: Lapland Chamber of Commerce. Accessed on 2 April 2021 from http://arcticbusinessforum.com/wp/wp-content/uploads/2019/09/ABF2019_yearbook.pdf

Royal Greenland. (2019). *Annual Report Royal Greenland A/S*. Accessed on 9 July 2020 from www.royalgreenland.com/globalassets/annual-report-2018/annual-report-2018.pdf

Royal Greenland. (n.d.a). *Royal Greenland – Corporate Presentation*. Accessed on 3 April 2021 from https://naalakkersuisut.gl/~/media/Nanoq/Files/Attached%20Files/Erhverv/Seoul_2016/7%20Royal%20Greenland%20Finn%20Lauersen.pdf

Royal Greenland. (n.d.b). Mission & vision. Accessed on 9 July 2020 from www.royalgreenland.com/royal-greenland/about-royal-greenland/mission--vision/

Royal Greenland. (n.d.c). Our approach to sustainability. Accessed on 9 July 2020 from www.royalgreenland.com/royal-greenland/sustainability/our-approach-to-sustainability/

Royal Greenland. (n.d.d). Our heritage. Accessed on 9 July 2020 from www.royalgreenland.com/royal-greenland/about-royal-greenland/our-heritage/

Royal Greenland. (n.d.e). Related documents. Sustainability in details – download area. Accessed on 9 July 2020 from www.royalgreenland.com/royal-greenland/sustainability/related-documents/

State of Alaska. (2019). EPA toxics release inventory not a reflection of risk. Accessed on 13 July 2020 from https://dec.alaska.gov/commish/press-releases/19-01-epa-toxics-release-inventory-not-a-reflection-of-risk/

Statistics Iceland. (2020). Iceland's population grew by 2.0% last year. Accessed on 15 May 2020 from www.statice.is/publications/news-archive/inhabitants/the-icelandic-population-1-january-2020-11643/

Teck. (2020a). *COVID-19 Testing and Travel Protocols for Red Dog Employees Traveling within the NANA Region*. Accessed on 13 July 2020 from www.teck.com/media/May-18-Red-Dog-Regional-Travel-Protocols.pdf

Teck. (2020b). *Red Dog COVID-19 Updated Travel Measures*. Red Dog Operations. Accessed on 13 July 2020 from www.teck.com/media/Red-Dog-COVID-19-Updated-Travel-Measures.pdf

Teck. (2017). *Red Dog Operations*. Accessed on 13 July 2020 from www.teck.com/media/P5-Red-Dog-Mine.pdf

Teck. (n.d.a). Responsibility. Accessed on 13 July 2020 from www.teck.com/responsibility/

Teck. (n.d.b). Sustainability report and disclosure portal. Accessed on 13 July 2020 from www.teck.com/responsibility/approach-to-responsibility/sustainability-report-disclosure-portal/

Tentipi. (2020). Inventing Stratus™: A tale of serendipity and hard work. Accessed on 19 June 2020 from https://tentipi.com/inventing-stratus,en-gb

Tentipi. (2019, January 16). 30 years crafting Nordic tipis. Accessed on 10 June 2020 from https://tentipi.com/en-gb/blog/30-years-crafting-nordic-tipis

Tentipi. (2019, April 18). 30th birthday celebrations in Arvidsjaur, Swedish Lapland. Accessed on 10 June 2020 from https://tentipi.com/en-gb/blog/30th-birthday-celebrations-in-arvidsjaur-swedish

Tentipi. (n.d.). Inspired by the Sami kåta. Accessed on 19 June 2020 from https://tentipi.com/event-our-nordic-roots,en-gb

United Nations Global Compact. (2020). 2018 Teck communication on progress. Accessed on 13 July 2020 from www.unglobalcompact.org/participation/report/cop/create-and-submit/advanced/435480

United Nations Global Compact. (n.d.). Transparency builds trust. Accessed on 13 July 2020 from www.unglobalcompact.org/participation/report/cop

West Baffin Eskimo Co-operative (WBEC). (n.d.a). A history of the WBEC. Accessed on 10 July 2020 from www.westbaffin.com/new-page-1

West Baffin Eskimo Co-operative (WBEC). (n.d.b). Home. Accessed on 10 July 2020 from www.westbaffin.com/

West Baffin Eskimo Co-operative (WBEC). (n.d.c). Kinngait studios. Accessed on 10 July 2020 from www.westbaffin.com/kinngait-studios

5 The role of business and stakeholders in the new CSR strategy in the Arctic

Learning objectives

- Identify what role businesses have in the new CSR strategy in the Arctic.
- Identify what role stakeholders have in the new CSR strategy in the Arctic.
- Identify the role of non-Arctic stakeholders in the new CSR strategy in the Arctic.
- Recognize the role of the environment and non-human stakeholders in the new CSR strategy in the Arctic.
- Recognize what distinguishes the Arctic stakeholder theory from contemporary stakeholder theory.
- Understand the importance of Arctic business principles and standards concerning the new CSR strategy in the Arctic.

Introduction

Businesses must have a vision about the role they play in managing change and exercise leadership in creating socially, economic, and environmentally sustainable development in the Arctic. This is achieved through a focus on the social responsibilities of companies and innovative partnerships between the public and private sectors, and civil society. The new perspective of higher and stricter socio-environmental standards to operate in the Arctic seems to be an important trend that requires not only formal regulation but also a serious commitment on the part of Arctic stakeholders, such as through commitment to corporate accountability principles.

Internal or external drivers drive organizations towards sustainability and socially responsible actions (e.g. Arena et al., 2018; Carroll and Shabana, 2010; Lozano, 2015). The internal drivers have, for instance, been categorized as ownership of companies; financial benefits of reducing costs, increasing sales, or a combination of both; or the ethical motivation of owners or managers. The external drivers pressuring companies to act responsibly include government and regulatory pressure, market pressure, social pressure, financial pressure, and pressure from the natural environment (Hart, 1995;

Bansal and Roth, 2000; Hoffman, 2000; Johannsdottir, 2015; Lozano, 2015). This suggests that companies act under pressure from a broad group of stakeholders, emphasizing the importance of exploring stakeholder theory and the role of stakeholders regarding the new corporate social responsibility (CSR) strategy in the Arctic.

Stakeholder theory has been used to emphasize organizational relationships and communication with actors within and outside organizations' boundaries; these actors have stakes in the activities organizations undertake. Therefore, stakeholder theory also emphasizes that organizations need to manage stakeholder expectations and reactions for their benefit (Bhattacharya et al., 2011). More recently, stakeholder theory also claims not only that businesses are accountable for their actions, but that stakeholders, such as regulators, employees, or consumers, are responsible for actions relevant for holding organizations accountable for their actions and therefore shape how they conduct their businesses (Chandler, 2017). Consequently, "*all stakeholders* have an obligation to help design the society in which they want to live and work" (Chandler, 2017, p. 93, emphasis in original). Therefore, organizations and stakeholders can be viewed either as independent or integrated actors, to the extent where more responsibility is placed on the shoulders of stakeholders than organizations (Chandler, 2017).

Types of discussion and issues explored in Chapter 5

This section provides a short overview of stakeholder theory and identifies the role of businesses and stakeholders in the new CSR strategy in the Arctic. The section also emphasizes the role of non-Arctic stakeholders, such as non-Arctic states and organizations, claiming a stake in the Arctic and its development. Given different Arctic conditions, the section highlights the role of the natural environment and non-human stakeholders in the new CSR strategy in the Arctic. This section also discusses what distinguishes contemporary stakeholder theory from Arctic stakeholder theory: the former focuses solely on stakeholders within the market economy, while Arctic stakeholder theory also recognizes stakeholders in mixed and subsistence economies and the environment and non-human stakeholders. This highlights the importance of strict socio-environmental Arctic business standards concerning the new CSR strategy in the Arctic and the importance of stakeholder dialogue if businesses are to conduct their activities in a responsible manner in the region.

Background: Stakeholder theory

Defining Arctic stakeholders is challenging, as both humans and non-humans have a stake in Arctic development, and both actors and peoples within and outside the region state their claim. One way to define immediate stakeholders is to include "the countries and organizations associated with the Arctic

Council", and those on whose behalf they represent interests and stakes; it is also stated that "we are all stakeholders in the Arctic" (Pálsson, 2003).

Role of business–stakeholder relationship in the new CSR strategy in the Arctic

CSR strategy will never be successfully implemented without stakeholders. Therefore, CSR scholars have been supportive of stakeholder theory (Freeman et al., 2017) where a stakeholder, by definition, is any group or individual who can affect or is affected by the achievement of the organization's objectives (Freeman, 1984). More specifically, stakeholders can be seen as "those groups without whose support the organization would cease to exist" (Freeman, 1984, p. 46). Stakeholder theory definitions have been categorized as claimant definitions, influencer definitions, and combinatory definitions (Kaler, 2002, pp. 94–96). According to the first type of definition, claimants are any group or individual "with a role-specific, strong or weak, morally legitimate claim to have their interests served". Influencers are defined as individuals or groups that are influenced or can influence the organization. Combinatory definitions include elements of both claimant and influencer definitions (Kaler, 2002). Furthermore, stakeholders can either voluntarily or involuntarily contribute to the wealth-creating activities and capacity of firms, and will as such either benefit or bear the risk of their actions (Post et al., 2002). In the case where the relationship is involuntary, stakeholders do not choose to enter into a relationship with a company but can at the same time not withdraw the stake they have in the firm (Clarkson, 1995).

Given the importance of stakeholder theory for CSR strategy and its implementation, the role of business in the new CSR strategy in the Arctic cannot be discussed without considering stakeholder theory and the relationship between businesses and their stakeholders. Contemporary stakeholder theory, which is a theory of organizational management and business ethics, is credited to R. Edward Freeman. At its core, the theory suggests that firms should create value for all stakeholders, not just shareholders (Freeman, 1984). Hopkins (2016) stated that these are "stakeholders of a company", such as employees, customers, suppliers, community and shareholders, and that "being a stakeholder is a matter of relationship" (p. 131). The primary purpose of the development of the theory was to solve three main problems: 1) to explain how value is created and trade made possible in an unstable business environment; 2) to comprehend the ethics of capitalism, and 3) what should be taught in business schools (Freeman et al., 2017).

There are, however, many ways to define stakeholders and the business–stakeholder relationship. One way is to categorize stakeholders as primary and secondary, whereas those who have a direct interest in the company, its activities, and success are primary and those who have indirect interests and are less directly connected to the core business are secondary. In generic terms, primary stakeholders include employees, customers, suppliers, financiers, and communities, while secondary stakeholders are for instance consumer advocate groups, competitors, government, media, special interest groups (Freeman et al., 2017), regulators, authorities, opinion formers and campaign groups, NGOs, religious institutions, and others (Hopkins, 2016).

Another grouping of stakeholders includes three groups. These are organizational stakeholders (employees, executives, and directors), economic stakeholders (customers, competitors, creditors, distributors/suppliers, shareholders, and unions), and societal stakeholders (communities, government agencies, media, non-profits/NGOs) (Chandler, 2017). There is an overlap between these two types of categorization, where most of the organizational and the economic stakeholders would be considered primary, but all of the societal stakeholders, except communities, would be classified as secondary. This categorization also groups stakeholders together instead of viewing them as individuals (Bhattacharya et al., 2011), but it should be noted that they are not all heterogenous. Figure 5.1 presents a synthesized stakeholder diagram (Bhattacharya et al., 2011; Hopkins, 2016; Freeman et al., 2017), presenting primary, secondary, organizational, economic, and societal stakeholders. Primary stakeholders are represented in dark grey boxes while secondary stakeholders are categorized according to organizational (light grey), economic (dark grey), and societal (medium grey) groups of stakeholders.

It is, furthermore, suggested that stakeholder want "a deeper and ongoing dialogue than that afforded by detailed annual reports", meaning that companies need to have a clear understanding of what stakeholders expect, what features they look for when they evaluate companies' CSR activities, and what their reactions will be (Bhattacharya et al., 2011, p. 76). Companies also need to understand that their stakeholders will not just rely on information provided by them, but will also obtain information from other sources and communication channels. Therefore, the information provided by businesses has to be clear, coordinated, and genuine, and CSR strategies need to be co-created, not developed through a top-down process in the company (Bhattacharya et al., 2011). The context of the relationship is also significant, such whether there are few or many stakeholders, whether the timeframe of the relationship is short or long, and whether the relationship is obligatory or whether it provides opportunities (Bhattacharya et al., 2011). The company's context explains issues such as the CSR programme of the company and how it is communicated, what its reputation is, and what is the profile of the company, ranging from small, local, and family-owned to multinational organizations. The stakeholder context explains how much stakeholders care about CSR issues, how close and embedded stakeholders are in the organization, and what their characteristics are (Bhattacharya et al., 2011).

Various models of stakeholder management exist, such as the synthesized model presented by Fifka and Loza Aduai (2015), based on Preble (2005); Buyssse and Verbeke (2003); Simmons and Lovegrove (2005); and Fassin (2009), see Figure 5.2. This six-step model consists of the following steps: Step

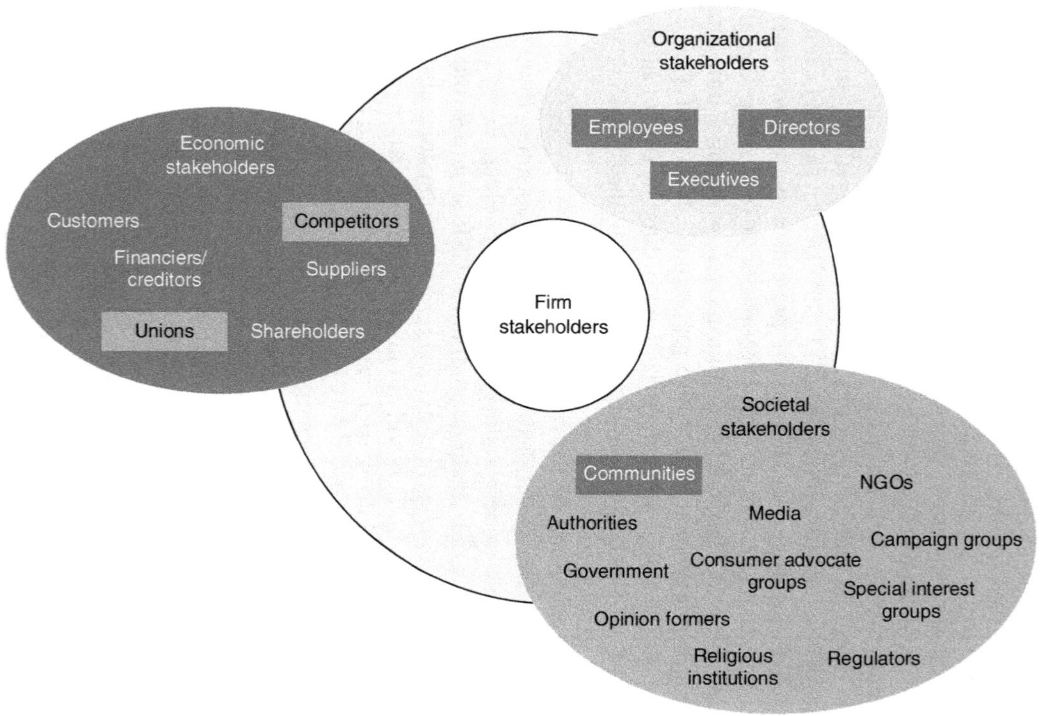

Figure 5.1 Synthesized stakeholder diagram

Figure 5.1 explains how organizational stakeholders can be grouped together, based on similar interests and influence, as organizational stakeholders, economic stakeholders, and societal stakeholders.

Source: Bhattacharya et al. (2011); Hopkins (2016); Freeman et al. (2017)

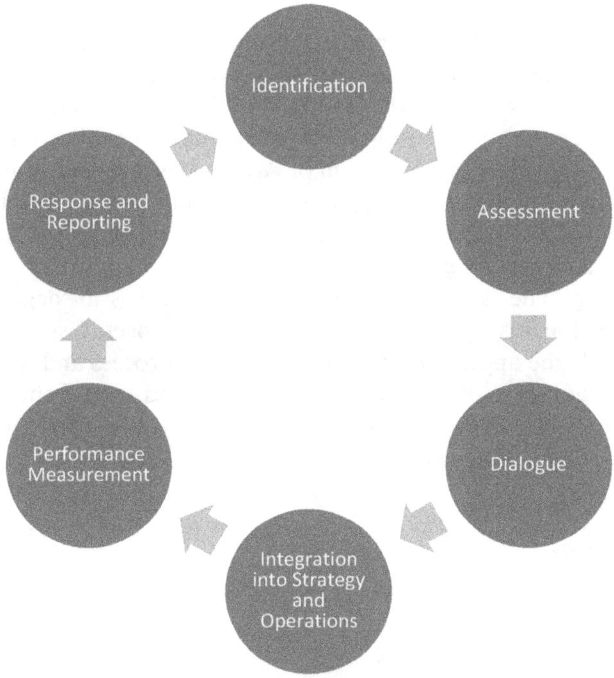

Figure 5.2 Stakeholder manager model

The stakeholder management model, presented in Figure 5.2, presents an example of how organizations can manage stakeholder interests and dialogue. This is presented in six steps: identification, assessment, dialogue, integration, performance measurements, and response and reporting.

Source: Fifka and Loza Aduai (2015, p. 79)

1. Identifying and listing stakeholders. Step 2. Assessing stakeholders. This is done to narrow the number of stakeholders and select those with whom to hold dialogue. Step 3. Dialogue. During this step, companies engage in a dialogue with the stakeholders so that their claims and interests can be identified. Step 4. Integration into strategy and operations. The information collected during Step 3 needs to be interpreted and issues crucial for the stakeholders and the company interpreted so that integration into strategy and operations can take place. Step 5. Performance measurement. Here previous actions are analysed and assessed against set goals and indicators. However, it should be noted that it is more challenging to assess environmental and social performance (Schaltegger and Wagner, 2004; Gjolberg, 2009) than financial performance. Step 5. Response and reporting. This is where the company communicates to stakeholders on what it has done to address their claims and interests. In general, the most impactful activities of companies are those that provide shared values of societal and economic progress (Porter and

Kramer, 2011) or "compelling benefits to the stakeholders" (Bhattacharya et al., 2011, p. 48).

Different methods for stakeholder engagement and involvement have been proposed, such as for issues surrounding the disposal of radioactive waste, but these methods are also relevant for other types of dialogue, engagement, and involvement. Many such examples exist in the Arctic context, including the case of dialogue with the Caribou Hunting System of Arctic Alaska where "hunting regulations in Alaska are informed through an open public process involving participation from different stakeholders" (Leorna, 2019, p iii), and dialogue used in relation to Arctic oil and gas development, where the aim was to shape mutual understanding. The latter cases highlight that there is a "degree of uncertainty regarding the actual outcomes of the Dialogue process", suggesting that an "effective dialogue appears to be a relatively slow process and it takes time to establish relationships on mutual understanding and trust from interpersonal level (participants) to inter-organisational level" (Makki, 2017, p. 41). Common methods include (Nuclear Energy Agency (NEA, 2004):

- Public hearings
- Deliberative polling
- Focus groups
- Citizen advisory groups
- Consultative groups
- Nominal group process
- Multi-actor policy workshops
- Charrette
- Delphi process
- Round tables
- Citizen task forces
- Study circles
- Scenario workshop
- Referendum
- Consensus conferences
- Citizen juries
- Citizen panels
- Participatory site selection

Engagement and involvement may take place in person, or communication can be virtual where the Internet and social media now play an important role. Stakeholders affected or those that can affect companies actively follow companies' activities and discussion of them, e.g. via news coverage, and provide feedback, positive or negative, on the coverage (Freeman et al., 2017). This is something companies need to be aware of, as in the case of a crisis where stakeholders' expectations are violated there may be negative outcomes for the organization and/or stakeholders (Coombs, 2017). Bad news about crisis, e.g. oil spills (Arruda, 2014, 2015; Johannsdottir and Cook, 2019), will spread around a region or the globe faster than a flu virus, with unpredictable consequences.

The integration of stakeholders' views in companies' decisions and actions can be defined as a "particular strategy of obtaining a competitive advantage through the development of close-knit ties with a broad range of internal and external constituencies" (Heugens et al., 2002, p. 38). Balancing the different claims of different stakeholders can be very challenging (Glomsrød et al., 2015). Therefore, as a part of stakeholder management, dialogue, engagement, and involvement it is suggested that companies recognize stakeholders' level of interest and power (Mitchell et al., 1997). If the levels of interest and power are low, then stakeholders should be monitored. If the level of interest is high, but their level of power is low, they should be informed. In the case where stakeholders' level of interest is low, but their level of power is high, they should be kept satisfied. However, if their level of interest and power are high, in both cases stakeholders should be managed closely. Stakeholder salience also adds another dimension to stakeholder management. This is the dimension of urgency in a matter; the urgency, legitimacy, and power of a stakeholder can be viewed narrowly from a management perspective (Mitchell et al., 1997), or more broadly where stakeholder salience is viewed as the relationship between stakeholders, firms, and their broader environment (Tashman and Raelin, 2013). This relationship has been tested in the case of two mining accidents in Canada and their impacts on gold mining stocks, with mixed results. The stock price did not decline after the first accident but did so after the second one, where shareholders' response became more profound (Magness, 2008).

Contemporary stakeholder theory is very market oriented. According to the theory, stakeholders are often viewed in relation to consumption, evident in their willingness to pay (or punish) companies, positive word-of-mouth, loyalty, or the employer–employee relationship is viewed in relation to recruitment, work performance, and turnover (Bhattacharya et al., 2011). In the Arctic context, the market view is too narrow, given that the market economy coexists with mixed and subsistence economies (Vammen Larsen et al., 2019). In the case of the Arctic, other interests matter just as equally. These include local values, traditional environmental knowledge, and priorities, based on centuries-long traditions and respect for the wisdom of the elders (CAFF, 2017; Saami Council, 2019) and those not yet born. These are becoming influential in research, policy, and management, as Indigenous peoples have to be consulted or directly involved in decision-making where decisions are made about their lands and activities through cooperative mechanisms, created by industry, to include all stakeholders (Glomsrød et al., 2015). The protection of Saami People's rights in EU investments has also been emphasized, and "free, prior and informed consent by the Saami communities impacted" by the activities has been ensured (Saami Council, 2019, p. 35).

Role of stakeholders in the new CSR strategy in the Arctic

CSR has been defined as "[a] responsibility among all of the firm's stakeholders to hold the firm to account for its actions by rewarding behaviour that meets expectations and punishes behaviour that does not" (Chandler, 2017, p. 94). It is furthermore stated that "stakeholders have an equal (if not more important) responsibility to hold firms to account from their actions" (Chandler, 2017, p. 71). The claim is that it suits for-profit firms better to react to economic motives and market forces than to determine trends in consumers' needs or market shifts. The role of stakeholders is consequently to "translate their values, needs, and concerns into action that punishes firms that fail to meet their criteria and rewards those that exceed it". Furthermore, it is suggested that "all stakeholders, not only consumers, have leverage over the firm and need to act" (Chandler, 2017, p. 94). Based on their interests, stakeholders define the parameters and boundaries for firms' operating conditions, and thus their success and profitability. The licence for them to operate therefore comes from society, not just authorities (Chandler, 2017).

In the role of holding organizations accountable, stakeholders can be caring, educated, informed, and/or transparent. In demonstrating stakeholder responsibility, caring stakeholders or activist stakeholders, have to care adequately about the CSR activities to permit corporate actions, value their legitimacy in society, or act in order to hold the organization accountable to protect their interest in its ongoing operations. Educated stakeholders take an ethical stance for, or against, organizational operations; this is grounded, for instance, in their education, such as CSR courses at university level, or business schools that use the UN's PRME Principles of Responsible Management Education as their founding principles for teaching. Communication technology supports stakeholders in their role as informed stakeholders, and actions can also be mobilized though such technology and platforms. Through various opinion surveys and ethics polls, stakeholders can be transparent about their concerns, needs, and values. Collectively, educated, informed, and transparent stakeholders that care are more engaged with their societies than those that do not fall under these categories. Thus, they are in a better position to shape the society they desire (Chandler, 2017). The claim is that stakeholder's reactions "as consumers, employees, investors, or even just the public – to a company's social responsibility actions (or the lack thereof) can have an unprecedented impact on its future" (Bhattacharya et al., 2011, p. 7). When defining those affected by organizations (voluntarily or involuntarily), the argument is that they pose the capacity of affecting the organization, and thus have the capacity to act in order to state their interests (Chandler, 2017). Non-acting entities that might not have recognized their self-interests, or cannot, are also recognized, such as infants, young children, and the environment. In such cases, other individuals or groups have to interpret their interests and act on behalf of those unable defend their interests.

Despite these theoretical claims, it can be hard and dangerous for stakeholders to hold companies accountable for their actions, or inactions. For instance, Indigenous communities often lack access to human capital and sustained investment (Glomsrød et al., 2015). Therefore, actors such as the Business & Human Rights Resource Centre (n.d.), a registered independent charity in the United Kingdom (UK) and a non-profit organization in the United States (USA), voice concerns of locals. It has regional researchers in Australia, Brazil, China, Colombia, India, Kenya, Jordan, Mexico, Myanmar, Philippines, Senegal, South Africa, Tunisia, the UK, Ukraine, and the US (Business & Human Rights Resource Centre, n.d.).

The challenge for stakeholders voicing their concerns over corporations' actions, such as human rights abuses and environmental issues, is that they sometimes face defamation lawsuits brought to shut down critics by tying them up in costly lawsuits, thus raising "barriers for victims to obtain justice" (Dowd & Aba, n.d.). The companies' actions are sometimes greatly exaggerated and disproportionate compared to the negative coverage of their actions (or inaction), which can even be as small as social media statements. Examples of exaggerated responses exist in Russia, where an Indigenous leader and member of the UN Working Group on Business and Human Rights, Pavel Sulyandziga and his family, have been harassed "because of his work supporting local communities to retain control of their land from extractives companies" (Dowd and Aba, n.d.). Even more serious is when those working on corporate accountability by defending human rights have faced beatings, threats, and even killings, and "are rarely, if ever, able to obtain justice". In two years, the Business & Human Rights Resource Centre tracked over 450 cases of attacks against these defenders of human rights. In most cases (40%), these are judicial harassments (Dowd and Aba, n.d.).

Stakeholders may also hold authorities or governments accountable for actions carried out by industries over a long period. This is evident in climate change complaints or litigation cases. For instance, in 2020 Louisiana and Alaskan tribes filed a formal complaint to the United Nations in Switzerland against the US government, stating that the government has failed to take action on climate change, thereby risking their homeland, identity, and culture (AP News, 2020). The government action, in this sense, would have been to set ambitious climate change goals, and using incentives, permits, taxes, and other means to reach the goals.

The media and environmental non-governmental organizations (ENGOs) can also voice the concern of local citizens, such as in oil spill cases in the Arctic. One such case is the 2020 major oil spill accident in the Russian Arctic, where an oil tank at the Nadezhda plant on the Taymyr peninsula, containing diesel oil, burst. The spill was in the excess of 20,000 metric tonnes of oil, resulting in contamination of more than 20km of the Siberian River (Greenpeace International, 2020). Such catastrophic incidents do not only require responses from the company, but have systematic consequences for local and federal governments and authorities, law enforcement bodies, and other stakeholders (Johannsdottir and Cook, 2019; CNN, 2020).

This suggests that when discussing the role of stakeholders in the new CSR strategy in the Arctic, one needs to understand that these stakeholders differ, as well as the stake they have in relation to corporate actions and inactions. However, their voices should not be silenced on the ground of different strengths, such as expertise and financial resources. Rather, corporations should find ways to collaborate with these stakeholders to bring about mutual benefits for both. Furthermore, different approaches can be used to promote greater accountability, such as in the case of respect for human rights. These include 1) binding treaties in cases where voluntary methods, such as CSR, are ineffective and lack enforcement mechanisms; 2) referencing guiding principles in legal proceedings, such as the Guiding Principles on Business and Human Rights; and 3) rating and ranking of companies' performance (Article One, 2017).

Role of non-Arctic stakeholders in the new CSR strategy in the Arctic

Those claiming a stake in the future of the Arctic and the economic development extend beyond the Arctic. This is evident in the observer status in the Arctic Council. This status is open to non-Arctic states, in conjunction with intergovernmental, inter-parliamentary, global, regional, and non-governmental organizations assuming they can contribute to the work of the Arctic Council. So far, 13 non-Arctic states have observer status in the Arctic Council; these include European and East and South Asian countries. Additionally, 13 intergovernmental and inter-parliamentary organizations have also gained observer status: for instance, the International Maritime Organization (IMO), the Nordic Council of Ministers (NCM), the United Nations Development Programme (UNDP), the United Nations Environment Programme (UNEP), and the World Meteorological Organization (WMO), just to name a few. Furthermore, 12 non-governmental organizations are approved observers in the Arctic Council. These consist, among others, of the Advisory Committee on Protection of the Sea (ACOPS), the International Arctic Science Committee (IASC), the International Arctic Social Sciences Association (IASSA), the International Work Group for Indigenous Affairs (IWGIA), and the World Wildlife Fund for Nature, Arctic Programme (WWF) (Arctic Council, n.d.).

International Arctic-related conferences also demonstrate a huge interest in the Arctic and Arctic affairs. The largest one is the Arctic Circle Assembly, which is an annual international gathering held in Reykjavík, Iceland, with over 2,000 participants from about 60 countries. Participants include head of states and high-level governmental officials, Indigenous Peoples, business representatives, researchers, NGOs and ENGOs, think-thanks, students, media, citizens, and other stakeholders (Arctic Circle, n.d.b). The purpose of the conference is to foster international dialogue and cooperation in the future of the Arctic. Participants have their own agenda for taking part in this platform, but they all have an interest in the development of the region, and what the consequences will be for the future of the globe (Arctic Circle, n.d.a). Other national, regional, or international conferences on Arctic affairs, held in different parts in the world, are smaller in terms of number of participants, but given the interest around the globe this emphasizes that Arctic development is not regarded as a private matter of the stakeholders living, regulating, or operating in the region.

The stakeholder interests in the Arctic are, furthermore, evident in regional and governmental policies, such as the European Union's (EU) Arctic policy, China and Japan's Arctic policies, and the Arctic policy of the Republic of Korea, just to name a few. The European Commission published an integrated EU policy for the Arctic in 2016, consisting of four categories: climate change and the Arctic environment, sustainable development in the Arctic, international cooperation on Arctic matters, and other matters (European Union External Actions, n.d). China's Arctic policy was published in a 2018 White Paper, containing five main policies and positions on China's participation in Arctic affairs: 1) deepening the exploration and understanding of the Arctic; 2) protecting the eco-environment of the Arctic and addressing climate change; 3) utilizing Arctic resources lawfully and rationally; 4) participating actively in Arctic governance and international cooperation; and 5) promoting peace and stability in the Arctic (State Council Information Office of the People's Republic of China, 2018). The outline of Japan's Arctic policy emphasizes research and development, international cooperation, and sustainable use of the Arctic sea route and mineral and living marine resources (Cabinet Office, n.d.). The vision of the Arctic policy of the Republic of Korea is to sustain the future of the Arctic, but the overarching policy goals are to build a cooperative Arctic partnership, enhance scientific research activities for the Arctic, and explore new business opportunities in the Arctic (Republic of Korea, n.d.).

Some of the main interests expressed by various States are synthesized in the EU's investment priorities in the Arctic. Through a consultation process involving EU institutions, Member States, and Arctic regional and local authorities, investment priorities were identified in relation to the digital infrastructure; internal and external transport connections; support of local business development, such as in renewable energy production, bioeconomy, sustainable use of natural resources and sustainable tourism; and climate change research, both mitigation and adaptation, and how opportunities which do not harm the environment and strengthen climate resilience can be utilized (European Commission, 2017). The role of non-Arctic stakeholders in the new CSR strategy in the Arctic is therefore vast, given that these stakeholders have both direct and indirect interest in the sustainable development of the region, and are in a position to influence responsible corporate conduct in the region.

Role of the environment and non-human stakeholders in the new CSR strategy in the Arctic

Most often, the stakeholder relationship is explained as the relationship between a corporation or an organization and actors within or outside their boundaries that can affect, or be affected by, their actions or inaction (e.g. Freeman, 1984). This is, however, a narrow view of the stakeholder concept, as it can also explain other types of relationships, such as between pets and their owners (Lauesen, 2013).

A narrow view of stakeholders is generally human-centric, although non-humans, such as the natural environment or animals, may have a stake in corporate actions and inaction, given the negative impacts human activities can have on non-human ecosystems and species (Woodward, 2002). The natural environment is sometimes presented as a stakeholder in stakeholder diagrams, such as in Figure 5.1. However, opinions are divided on whether to count the natural environment and non-human stakeholders as stakeholders or not. Some provide arguments for why they should not be (Hopkins, 2016; Chandler, 2017), such as that they do not feel, speak, or act, and thus lacks agency (Chandler, 2017). Logics for including the environment is that it can be hard to identify humans that may be affected by companies' actions, such as in the case of global warming impacts, environmental degradation, or pollution. Hence, the stakeholder group would be considered to be those who are affected by these issues (Chandler, 2017; Hopkins, 2016). The argument for not including the environment as stakeholders or in stakeholder diagrams, even though the importance of sustainability is emphasized, is that the environment requires societal actors to speak enthusiastically and knowledgeably on its behalf and act accordingly so that its stake is protected (Chandler, 2017); therefore, there is no need for including it per se. These entities therefore may need spokesperson(s) who can present their interests by representing them as stakeholders (Lauesen, 2013) for the benefit of coexistence between humans and non-humans. The claim is that "a company's long-term success, and sometimes existence, is inextricably tied to its stewardship of not just its own well-being but also that of the natural and social environment in which it operates" (Bhattacharya et al., 2011, p. 8).

This leads to a discussion about the principles of weak and strong sustainability. The former is grounded in economics with the claim that natural capital can subsidized by human capital, whereas the latter emerged from ecological economics where the idea is that economic and ecological capital are not interchangeable but complementary (Werner, 1999). Moreover, there is a claim that "an integrated approach is required that is based on an ecological-economic system perspective of development and preservation of the environment as our life-support system" (Werner, 1999, p. 1133). Development and progress are not without consequences and after the Second World War, some of the damages to the natural environment caused by chemical pollution became evident since it was of such scale that it could danger human health. This has resulted in legal restrictions, or bans, on using some chemicals (Tokuç, 2013), or international treaties such as the 2001 Stockholm Convention on Persistent Organic Pollutants that aims to eliminate or restrict the production and use of persistent organic pollutants (Secretariat of the Stockholm Convention, 2019).

Voices are getting louder that nature has its own rights. This includes animals, mountains, oceans, and trees that have rights just as human beings. These rights are honoured when the rights of humans and nature are balanced in such a way that what is good for humans is also good for other species and the planet as a whole. This perspective recognizes how deeply intertwined all ecosystems are and that nature and "all its life forms have the *rights to exist, persist, maintain and regenerate its vital cycles*" instead of being treated under property laws (Global Alliance for the Rights of Nature, n.d., emphasis in original). The first nation to recognize the rights of nature in its Constitution was Ecuador. In Chapter 7 of the Constitution, on the rights of nature, four articles contain messages of relevance for the chapter discussion (Republic of Ecuador, 2008):

> Article 71: All persons, communities, peoples and nations can call upon public authorities to enforce the rights of nature.
> Article 72: In those cases of severe or permanent environmental impact, including those caused by the exploitation of nonrenewable natural resources, the State shall establish the most effective mechanisms to achieve the restoration and shall adopt adequate measures to eliminate or mitigate harmful environmental consequences.
> Article 73: The State shall apply preventive and restrictive measures on activities that might lead to the extinction of species, the destruction of ecosystems and the permanent alteration of natural cycles.
> Article 74: Persons, communities, peoples, and nations shall have the right to benefit from the environment and the natural wealth enabling them to enjoy the good way of living.
> Environmental services shall not be subject to appropriation; their production, delivery, use and development shall be regulated by the State.

The role of the environment and non-human stakeholders in the new CSR strategy in the Arctic is extremely important, given centuries of close ties between the human and non-human worlds, evident in the livelihood, culture, and identity of those living and working in the region. For businesses, as well as other actors, supporting sustainable development, it is therefore important not to limit their view to a narrow stakeholder one; rather, they should view stakeholders from a much broader perspective, such as by taking into account the stakes of unborn children and non-human actors, such as animals and ecosystems (Remme, 2013). This is challenging, but nonetheless important when it comes to Arctic development and issues since such development is not just about the market economy, but also mixed and subsistence economies (Vammen Larsen et al., 2019) where there is great respect for coexistence between humans and non-humans.

Given the fragile Arctic environment and the coexistence between humans and non-humans in the region, the role of the environment and non-human stakeholders in the new CSR strategy in the Arctic is, therefore, of great importance, even though their stakes have to presented by others, such as through international treaties, laws, regulation, NGOs, or other representing their interests.

Arctic stakeholder theory versus contemporary stakeholder theory

Contemporary stakeholder theory is very human and market-centric (Woodward, 2002; Bhattacharya et al., 2011) and as such too narrow from the Arctic perspective where three types of economies co-exist: market, mixed, and subsistence economies (Vammen Larsen et al., 2019). Therefore, building on the synthesized stakeholder diagram (Bhattacharya et al., 2011; Hopkins, 2016; Freeman et al., 2017), Figure 5.1, a new diagram is proposed for firms operating responsibly in the Arctic region. It adds new dimensions to the former diagram and is, therefore, more complex but more descriptive of Arctic conditions. It takes into account the Arctic institutional framework, including the Arctic Council, local and regional laws and regulations, the Arctic Economic Council, various Indigenous and wildlife organizations and others (Vammen Larsen et al., 2019), human stakeholders (Lauesen, 2013; Chandler, 2017), in addition to the natural environment and other non-human stakeholders (Remme, 2013). The argument for excluding the natural environment from the stakeholder diagrams (Chandler, 2017) also arguably holds for other human stakeholders this figure recognizes, in the sense that some societal stakeholder should act on their behalf. This view involves a great deal of risk. What if no one takes on this role and responsibility, and high stakes are left unaddressed or unprotected? This argument is therefore considered unacceptable in the Arctic context where the human and non-human ecosystems are intertwined (Global Alliance for the Rights of Nature, n.d.). In this region, the context of the relationship should be considered and co-creation of CSR strategies is an essential element (Bhattacharya et al., 2011), such as the integration of the environmental, social, and cultural context that represents the homeland and identity of people living in the region, and the coexistence of different types of economies (Vammen Larsen et al., 2019).

Arctic business principles and standards concerning the new CSR strategy in the Arctic

The new perspective of higher and stricter socio-environmental standards operating in the Arctic seems to be an important trend that requires not only formal regulation but also a serious commitment on the part of Arctic stakeholders, such as through commitment to corporate accountability

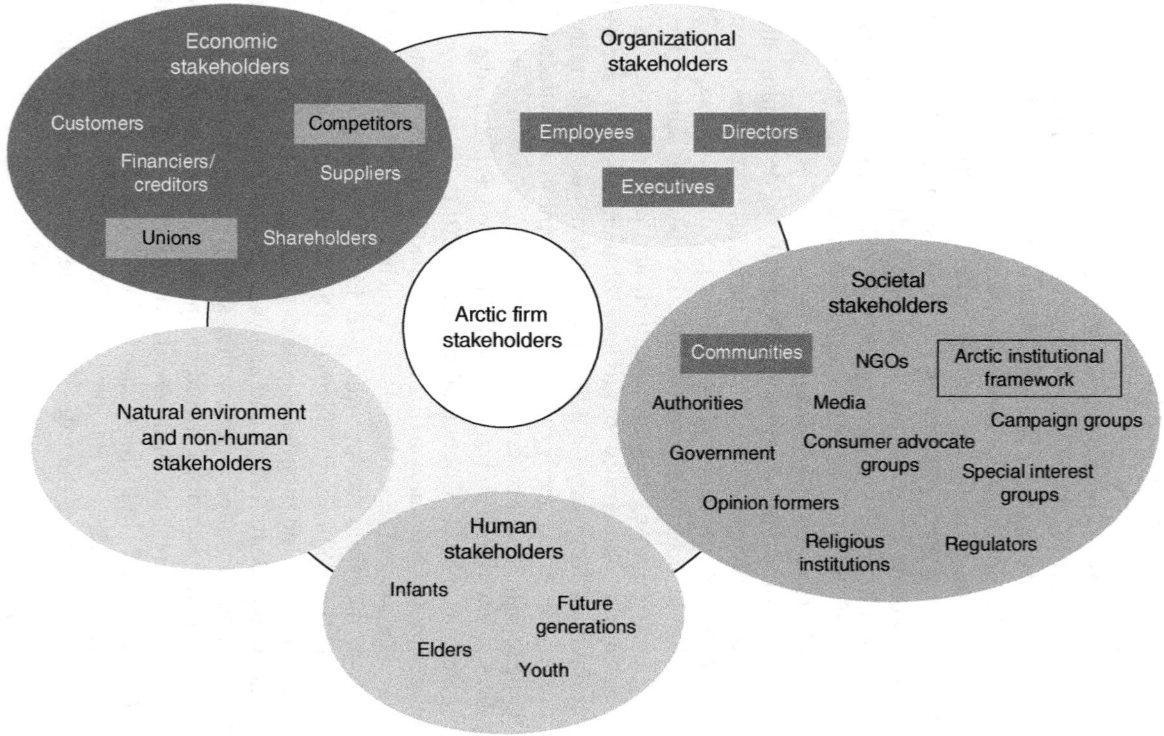

Figure 5.3 Stakeholder diagram for firms operating in the Arctic region

The stakeholder diagram for firms operating in the Arctic region includes common groups of stakeholders, namely organizational stakeholders, economic stakeholders, and societal stakeholders, but also adds two additional groups: human stakeholders, and the natural environment and non-human stakeholders.

Source: Adapted from Bhattacharya et al. (2011); Freeman et al. (2017); Hopkins (2016)

principles. International agreed standards, guidelines, principles, and initiatives play a crucial role in addressing social and environmental challenges related to business conduct, thus supporting higher and stricter socio-environmental standards. They also highlight growing expectations from stakeholders on how to institutionalize responsible conduct (Nordic Council of Ministers, 2012). Among a wide number and variety of initiatives that are complementary and influence responsible business conduct are the OECD Guidelines for Multinational Enterprises, the UN Guiding Principles on Business and Human Rights, the UN Principles for Responsible Investment, the UN Global Compact, the ISO 26000 Guidance for Social Responsibility, and the broader UN sustainability agenda (Nordic Council of Ministers, 2012). These are discussed in more detail in other chapters of the book, such as in Chapter 9. In carrying out responsible business conduct in the Arctic, development of concrete tools and best practices in consultation with relevant stakeholders is emphasized, and use of the OECD Guidelines for Multinational Enterprises recommended (Nordic Council of Ministers, 2012).

While companies should have a degree of flexibility in developing their CSR approach in accordance with their circumstances, using standards levels the playing field between companies (Nordic Council of Ministers, 2012). Consequently, it should lead to comparability that benefits stakeholders when evaluating and comparing business conduct, and thus promote trust. Furthermore, given the complexity of today's challenges, it is deemed essential for companies to engage stakeholders in dialogue so that societal expectations and business needs can be balanced, and issues effectively addressed. Dialogue with stakeholders is becoming mainstream and critical for the integration of environmental, social, and ethical concerns into day-to-day business operations (Nordic Council of Ministers, 2012).

Another suitable tool of relevance in the Arctic context is the Environmental and Social Impact Assessment (ESIA) guidance, which is applicable to projects that involve high or moderate risk according to Environmental and Social Management System (ESMS) screening. This tool helps identify adverse social and environmental impacts and propose measures to mitigate these impacts. It also includes stakeholder identification and analysis, thus clarifying those that need to be involved and consulted and how they should be involved. It offers a way to analyse alternative solutions, including an option where the project is not recommended, based on the potential severity of risk events and likelihood of occurrence (International Union for Conservation of Nature, 2020).

More demanding are the 10 Principles for Corporate Accountability, proposed by Greenpeace, where the role of the State in overseeing business actions and in protecting environmental and human rights are emphasized (Greenpeace, 2018, p. 9).

1. People and the environment, not corporations, must be at the heart of governance and public life.
2. Public participation should be inherent to all policy-making.

3. States should abandon policies that undermine environmental and human rights.
4. Corporations should be subject to binding rules both where they are based and where they operate.
5. States should require due diligence reporting and cradle to grave responsibility for corporate products and services.
6. States should promote a race to the top by prohibiting corporations from carrying out activities abroad which are prohibited in their home state for reasons of risks to environmental or human rights.
7. States should create policies that provide transparency in all corporate and government activities that impact environmental and human rights, including in trade, tax, finance and investment regimes.
8. Corporations and those individuals who direct them should be liable for environmental and human rights violations committed domestically or abroad by companies under their control.
9. People affected by environmental and human rights violations should be guaranteed their right to effective access to remedy, including in company home states where necessary.
10. States must actually enforce the regulatory and policy frameworks they create.

The 10 Principles for Corporate Accountability set by Greenpeace make greater demands on States than businesses, which makes them useful in parallel with other principles, standards, guidelines, and initiatives, presented in this section.

Further resources

CSR, and responsible business conduct, is carried out by business leaders who are knowledgeable and respectful of the circumstances in which their businesses operate, including the inclusion of relevant stakeholders. The Principles of Responsible Management Education (PRME), a UN-supported initiative founded in 2007, lays the foundation for such practices. The purpose of this platform is to strengthen sustainability emphasis in business schools around the globe so they are equipped to understand and can address tomorrow's challenges. Business schools have committed to the six principles, but the purpose is to shape the mindset and skills of future leaders (PRME, n.d. a, b), all of which are relevant for business leaders conducting business in the Arctic region.

Principle 1: Purpose. We will develop the capabilities of students to be future generators of sustainable value for business and society at large and to work for an inclusive and sustainable global economy.

Principle 2: Values. We will incorporate into our academic activities, curricula, and organisational practices the values of global social responsibility as portrayed in international initiatives such as the United Nations Global Compact.

Principle 3: Method. We will create educational frameworks, materials, processes, and environments that enable effective learning experiences for responsible leadership.

Principle 4: Research. We will engage in conceptual and empirical research that advances our understanding about the role, dynamics, and impact of corporations in the creation of sustainable social, environmental, and economic value.

Principle 5: Partnership. We will interact with managers of business corporations to extend our knowledge of their challenges in meeting social and environmental responsibilities and to explore jointly effective approaches to meeting these challenges.

Principle 6. Dialogue. We will facilitate and support dialog and debate among educators, students, business, government, consumers, media, civil society organisations and other interested groups and stakeholders on critical issues related to global social responsibility and sustainability.

Case study

Business Index North (BIN) is a project with the aim of contributing to sustainable development and value creation in the Arctic through cooperation and future development. Its "overall goal is to set up a recurring, knowledge-based, systematic information tool for stakeholders" (Middleton et al., 2020, p. 32). What has been noted through this project is that many stakeholders lack an overview of entrepreneurial activities taking place in the region (Middleton et al., 2020). The High North Center for Business and Governance at Nord University Business School (Norway) coordinates the BIN project, which is implemented through an international network of partners from the northern regions of Finland, Sweden, Norway, and North-West Russia (Middleton et al., 2019). The future aim is to include "Alaska and the Northern territories of Canada, Iceland, and Greenland, as well as more territories of the Russian High North" (Middleton et al., 2019, p. 52).

BIN reports, titled *Business Index North*, are published regularly, and "serve as an analytical tool for various stakeholders, including decision-makers, media, NGOs, academia and others" (Middleton et al., 2019, p. 51). Through the statistical analysis, list of companies and their activities, and other relevant information, these reports include information about the Arctic people, businesses, and development conditions (Middleton et al., 2019, p. 1):

- People: Provides analysis of demographic and human capital trends in the region.
- Business: Gives an overview of business activity and perspective of value creation.
- Development conditions: Focuses on key issues of connectivity, R&D in business, and electricity production.

Questions for discussion

1. Who are Arctic stakeholders?
2. What stake do they have the in the development of the Arctic region?
3. Why do some stakeholders involuntarily contribute to the wealth-creating activities and capacity of firms?
4. Who can voice the concerns of those unable to defend their own stakes?
5. What is the difference between contemporary stakeholder theory and Arctic stakeholder theory?
6. What does it mean if the information level and bargaining power between businesses and stakeholders are unbalanced, and how can this issue be mitigated?
7. Why are Arctic business principles and standards concerning the new CSR strategy in the Arctic important?

Chapter summary

A prerequisite in the new CSR strategy in the Arctic, and to enable flourishing and thriving communities, businesses, and natural environment, is the inclusion of stakeholders, both human and non-human. This suggests that the drivers of sustainability for organizations need to be both to internal and external. Internal drivers include ethical motivation while external drivers (e.g. Carroll and Shabana, 2010; Lozano, 2015; Arena et al., 2018) include pressure from outside actors, whether governmental bodies, media, or NGOs (Hart, 1995; Bansal and Roth, 2000; Hoffman, 2000; Johannsdottir, 2015; Lozano, 2015), or CSR-related principles, standards, guidelines, and other such means. This chapter provided an overview of stakeholder theory, identifying the role of businesses and stakeholders in the new CSR strategy in the Arctic. It also emphasized the role of non-Arctic stakeholders, such as non-Arctic states and organizations claiming a stake in the Arctic and its development.

The role of the natural environment and non-human stakeholders in the new CSR strategy in the Arctic was furthermore highlighted in this chapter. A distinction was made between contemporary stakeholder theory (Freeman, 1984) and Arctic stakeholder theory, explained through two diagrams, Figures 5.1 and 5.3, respectively. The former focuses solely on stakeholders within the market economy, while Arctic stakeholder theory also recognizes stakeholders in mixed and subsistence economies and the importance of environmental and non-human stakeholders, although they are unable to voice their concerns and defend their own stakes. It is perhaps more important for business leaders conducting business in the region to understand these conditions than categorizing stakeholders as primary, secondary (Freeman et al., 2017), organizational, economic, or societal stakeholders (Chandler, 2017). They need to understand that stakeholders sometimes involuntarily contribute to the wealth-creating activities and capacity of firms by bearing the risk of their actions (Post et al., 2002), and therefore are not able to withdraw the stake they have in the firm (Clarkson, 1995). Furthermore, it should be recognized that in the business–stakeholder relationship there is often a very different information level and imbalance in bargaining power given the financial strengths of corporations presenting large-scale commercial interests. Therefore, although stakeholders may have a more important role to play in holding business accountable for their actions rather than the stakeholders themselves, and therefore shape how they conduct their businesses (Chandler, 2017), this may not be the case in the Arctic business–stakeholder relationship. Consequently, this highlights the importance of strict socio-environmental Arctic business standards in the new CSR strategy in the Arctic and the importance of stakeholder dialogue if businesses are to conduct their activities responsibly in the region. Responsible business conduct in the region, where the rights of humans and nature are respected, will benefit businesses through gaining the social license to operate. People's stakes, and the stake of businesses, are inherently tied together.

References

AP News. (2020, January 18). Louisiana, Alaskan tribes file UN climate change complaints. Accessed on 15 July 2020 from https://apnews.com/b324677e542281b23d811554dba6cd79

Arctic Circle. (n.d.a). About. Accessed on 16 July 2020 from www.arcticcircle.org/about/about/

Arctic Circle. (n.d.b). 2019 Assembly. Accessed on 16 July 2020 from www.arcticcircle.org/

Arctic Council. (n.d.). Observers. Accessed on 16 July 2020 from https://arctic-council.org/en/about/observers/

Arena, M., Azzone, G., and Mapelli, F. (2018). What drives the evolution of Corporate Social Responsibility strategies? An institutional logics perspective. *Journal of Cleaner Production*, 171, 345–355. https://doi.org/10.1016/j.jclepro.2017.09.245

Arruda, G.M. (2014). Global governance, health systems and oil and gas exploration. *International Journal of Law and Management*, 56(6), 495–508.

Arruda, G.M. (2015). Arctic governance regime: The last frontier for hydrocarbons exploitation, *International Journal of Law and Management*, 57(5), 498–521.

Article One. (2017). Three ways stakeholders hold companies accountable for human rights. Accessed on 15 July 2020 from www.articleoneadvisors.com/insights/2017/1/18/corporate-accountability

Bansal, P. and Roth, K. (2000). Why companies go green: A model of ecological responsiveness. *Academy of Management Journal*, 43(4), 717e736. http://dx.doi.org/10.2307/1556363.

Bhattacharya, C.B., Sen, S., and Korschun, D. (2011). *Leveraging Corporate Responsibility. The Stakeholder Route to Maximizing Business and Social Value*. Cambridge, UK: Cambridge University Press.

Business & Human Rights Resource Centre. (n.d.). Who and where we are. Accessed on 15 July 2020 from www.business-humanrights.org/en/about-us/who-and-where-we-are

Buysse, K. and Verbeke, A. (2003). Proactive environmental strategies: A stakeholder management perspective. *Strategic Management Journal*, 24(5), 453–470.

Cabinet Office. (n.d.). *Outline of Japan's Arctic Policy*. Accessed on 16 July 2020 from www8.cao.go.jp/ocean/english/arctic/pdf/outline_japans_ap_e.pdf

Carroll, A.B. and Shabana, K.M. (2010). The business case for corporate social responsibility: A review of concepts, research and practice. *International Journal of Management Reviews*, 12(1), 85–105. doi:10.1111/j.1468-2370.2009.00275.x

Chandler, D. (2017). *Strategic Corporate Social Responsibility: Sustainable Value Creation* (4th ed). Thousand Oaks, CA: Sage.

Clarkson, M. (1995). A stakeholder framework for analysing and evaluating corporate social performance. *Academy of Management Review*, 20(1), 92–117.

CNN. (2020, June 4). Putin declares emergency over 20,000 ton diesel spill. Accessed on 15 July 2020 from https://edition.cnn.com/2020/06/03/europe/russia-putin-oil-spill-norilsk-intl/index.html

Conservation of Arctic Flora and Fauna (CAFF). (2017). Arctic Traditional Knowledge and Wisdom: Changes in the North American Arctic. *Arctic Council Perspectives from Arctic Athabascan Council, Aleut International Association, Gwich'in Council International, and Published Accounts* (CAFF Assessment Series Report No. 14). Akureyri, Iceland: Author.

Coombs, T. (2017). CSR and crisis communication strategies. In A. Rasche, M. Morsing, and J. Moon (Eds.), *Corporate Social Responsibility. Strategy, Communication, Governance* (pp. 307–327). Cambridge, UK: Cambridge University Press.

Dowd, C. and Aba, E. (n.d.). Why it's getting harder (and more dangerous) to hold companies accountable. Accessed on 16 July 2020 from www.business-humanrights.org/en/why-it%E2%80%99s-getting-harder-and-more-dangerous-to-hold-companies-accountable-0

European Commission. (2017). *Summary Report of the Arctic Stakeholder Forum Consultation to Identify Key Investment Priorities in the Arctic and Ways to Better Streamline Future EU Funding Programmes for the Region*. Luxembourg: Publications Office of the European Union.

European Union External Actions. (n.d.). EU Arctic Policy. Accessed on 16 July 2020 from https://eeas.europa.eu/arctic-policy/eu-arctic-policy_en

Fassin, Y. (2009). The stakeholder model refined. *Journal of Business Ethics*, 84(1), 113–135.

Fifka, M. and Loza Aduai, C.R. (2015). Managing stakeholders for the stake of business and society. In L. Riordan, P. Zmuda, and S. Heinemann (Eds.), *New Perspective on Corporate Social Responsibility: Locating the Missing Link* (pp. 71–87). Wiesbaden, Germany: Springer Gabler.

Freeman, E.R., Dmytriyev, S., and Strand, R.G. (2017). Managing for stakeholders in the digital age. A. Rasche, M. Morsing, and J. Moon (Eds.), *Corporate Social Responsibility: Strategy, Communication, Governance* (pp. 136–153). Cambridge, UK: Cambridge University Press.

Freeman, R.E. (1984). *Strategic Management: A Stakeholder Approach*. Boston, MA: Pitman. (Reprinted in 2010, Cambridge University Press).

Gjolberg, M. (2009). Measuring the immeasurable? Constructing an index of CSR practices and CSR performance in 20 countries. *Scandinavian Journal of Management*, 25(1), 10–22.

Global Alliance for the Rights of Nature. (n.d.). What is Rights of Nature? Accessed on 20 July 2020 from https://therightsofnature.org/what-is-rights-of-nature/

Glomsrød, S., Duhaime, G., and Aslaksen, I. (Eds.). (2015). *The Economy of the North 2015*. Oslo–Kongsvinger, Norway: Statistics Norway.

Greenpeace. (2018). *Justice for People and Planet: Ending the Age of Corporate Capture, Collusion and Impunity*. Amsterdam: Author.

Greenpeace International (2020, June 5). The river runs red – catastrophic oil accident in the Russian Arctic. Accessed on 15 July 2020 from www.greenpeace.org/international/story/43553/oil-spill-russian-arctic/

Hart, S.L. (1995). A natural-resource-based view of the firm. *Academy of Management Review*, 20(4), 986e1014.

Heugens, P.P.M.A.R., van den Bosch, F.A.J., and van Riel, C.B.M. (2002). Stakeholder integration: Building mutually enforcing relationships. *Business & Society*, 41(1), 36–60.

Hoffman, A.J. (2000). *Competitive Environmental Strategy: A Guide to the Changing Business Landscape*. Washington, D.C.: Island Press.

Hopkins, M. (2016). *CSR and Sustainability. From the Margins to the Mainstream: A Textbook*. Sheffield: Greenleaf Publishing.

International Union for Conservation of Nature. (2020). *Environmental and Social Impact Assessment (ESIA)*, version 15 March 2020. Accessed on 20 July 2020 from www.iucn.org/sites/dev/files/iucn_esms_esia_guidance_note.pdf

Johannsdottir, L. (2015). Drives of proactive environmental actions of small, medium and large Nordic non-life insurance companies e and insurers as a driving force of actions. *Journal of Cleaner Production*, 108(2015), 685–698. http://dx.doi.org/10.1016/j.jclepro.2015.06.055

Johannsdottir, L. and Cook, D. (2020). Systemic risk of maritime-related oil spills viewed from an Arctic and insurance perspective. *Ocean & Costal Management*, 179(1), 104853. https://doi.org/10.1016/j.ocecoaman.2019.104853

Kaler, J. (2002). Morality and strategy in stakeholder identification. *Journal of Business Ethics*, 39(1--2), 91–100.

Lauesen. L.M. (2013). Stakeholder accountability. In S.O. Idowu, N. Capaldi, L. Zu, and A.D. Gupta (Eds.), *Encyclopedia of Corporate Social Responsibility*. Springer. https://doi.org/10.1007/978-3-642-28036-8_92

Leorna, S.S. (2019). *Exploring Stakeholder Communication within a Caribou Hunting System of Arctic Alaska* (unpublished MSc thesis). Alaska: University of Alaska Fairbanks.

Lozano, R. (2015). A holistic perspective on corporate sustainability drivers. *Corporate Social Responsibility and Environmental Management*, 22(1), 32–44. http://dx.doi.org/10.1002/csr.1325

Magness, V. (2008). Who are the stakeholders now? An empirical examination of the Mitchell, Agile, and Wood theory of stakeholder salience. *Journal of Business Ethics*, 83(2), 177–192.

Makki, M. (2017). Evaluating Arctic dialogue: A case study of stakeholder relations for sustainable oil and gas development. *Journal of Sustainable Development*, 8(1), 34–45. doi:10.5539/jsd.v5n3p34

Middleton, A., Hersinger, A., Bryksenkov, A., Mineev, A., Dybtsyna, E., Bullvåg, E., Simonen, J., Pesämaa, O., Dahlin, P., and Ovesen, S. (2019). *Business Index North*. A periodic report with insight to business activity and opportunities in the Arctic. Issue 3, vol. June 2019.

Middleton, A., Pesämaa, O., Dahlin, p., Mineev, A., and Bullvåg, E. (2020). Innovative businesses in the Arctic: Many ways to succeed. *Business Index North* (special issue, January 2020).

Mitchell, R.K., Agle, B.R., and Wood, D.J. (1997). Toward a theory of stakeholder identification and salience: Defining the principle of who and what really counts. *Academy of Management Review*, 22(4), 853–886.

Nordic Council of Ministers. (2012). *Nordic Strategy for Corporate Social Responsibility*. Copenhagen K, Denmark: Author.

Nuclear Energy Agency (NEA). (2004). *Stakeholder Involvement Techniques: Short Guide and Annotated Bibliography* (NEA/RWM/FSC(2004)7/PROV). Paris, France: OECD.

Pálsson, G. (2003). Norðurslóðir og umheimurinn. Speech given at the Greenwich Forum. Accessed on 20 July 2020 from www.stjornarradid.is/efst-a-baugi/frettir/stok-frett/?NewsID=144321f2-e40c-11e6-9c23-e4a47104df25

Porter, M.E. and Kramer, M.R. (2011). Creating shared value: How to reinvent capitalism—and unleash a wave of innovation and growth. *Harvard Business Review* (January–February 2011).

Post, J.E., Preston, L.E., and Sauter-Sachs, S. (2002). *Redefining the Corporation: Stakeholder Management and Organizational Wealth*. Palo Alto, CA: Stanford University Press.

Preble, J.F. (2005). Toward a comprehensive model of stakeholder management. *Business and Society Review*, 110(4), 407–431.

Principles for Responsible Management Education (PRME). (n.d.a). What is PRME? Accessed on 10 July 2020 from www.unprme.org/about

Principles for Responsible Management Education (PRME). (n.d.b). What we do? Six principles. Accessed on 20 July 2020 from www.unprme.org/what-we-do

Remme, J. H. M. (2013). Stakeholder thinking. In S. O. Idowu, N. Capaldi, L. Zu, and A.D. Gupta (Eds.), *Encyclopedia of Corporate Social Responsibility*. Springer. https://doi.org/10.1007/978-3-642-28036-8_92

Republic of Ecuador. (2008). Constitution of the Republic of Ecuador. Accessed on 20 July 2020 from https://pdba.georgetown.edu/Constitutions/Ecuador/english08.html

Republic of Korea. (n.d.). Arctic Policy of the Republic of Korea. Accessed on 16 July 2020 from http://koarc.org/theme/northpn_ko/03/pdf/korea_eng.pdf

Saami Council. (2019). *The Sámi Arctic Strategy*. Kárášjohka, Norway: Sámiráđđi.

Schaltegger, S. and Wagner, M. (2004). The effect of corporate environmental strategy choice and environmental performance on competitiveness and economic performance: An empirical study of EU manufacturing. *European Management Journal*, 22(5), 557–572. https://doi.org/10.1016/j.emj.2004.09.013

Secretariat of the Stockholm Convention. (2019). History of the negotiations of the Stockholm Convention. Accessed on 18 July 2020 from www.pops.int/TheConvention/Overview/History/Overview/tabid/3549/Default.aspx

Simmons, J. and Lovegrove, I. (2005). Bridging the conceptual divide: Lessons from stakeholder analysis. *Journal of Organizational Change Management*, 18(5), 495–513.

State Council Information Office of the People's Republic of China. (2018). *China's Arctic Policy* (1st ed.). Accessed on 16 July 2020 from http://english.www.gov.cn/archive/white_paper/2018/01/26/content_281476026660336.htm

Tashman, P. and Raelin, J. (2013). Who and what really matters in the firm: Moving stakeholder salience beyond managerial perceptions. *Business Ethics Quarterly*, 23(4), 591–616.

Tokuç, A. (2013). Stockholm Convention (2001). In S.O. Idowu, N. Capaldi, L. Zu, and A.D. Gupta (Eds.), *Encyclopedia of Corporate Social Responsibility*. Springer. https://doi.org/10.1007/978-3-642-28036-8_92

Vammen Larsen, S., Bors, E., Johannsdottir, L., Gladun, E., Gritsenko, D., Nysten-Haarala, S., Tulaeva, S., & Sformo, T. (2019). A conceptual framework of Arctic economies for policy-making, research, and practice. *Global Policy*, 20(4): 686–696. doi:10.1111/1758-5899.12720

Werner, H. (1999). Reconciling "weak" and "strong" sustainability. *International Journal of Social Economics*, 26(7–9), 1120–1144. doi:10.1108/03068299910245859

Woodward, D.G. (2002). Is the natural environment a stakeholder? Of course it is (no matter what the Utilitarians might say)! In *Critical Perspectives on Accounting Conference*, 25–27 April 2002. New York, Baruch College: City University of New York.

6 CSR and Arctic communities

Learning objectives

- Understand responsible development in the Arctic context.
- Understand the balance of economic, social, and environmental aspects of corporate operations.
- Understand traditional knowledge as a basis for business development of Arctic natural resources.
- Understand that socio-technical systems co-evolve.
- Understand the concept of a low-carbon Arctic smart community.
- Understand and evaluate the viability of Arctic agroecosystems.

Introduction

Corporate social responsibility (CSR) is the commitment on the part of businesses to behave ethically and contribute to economic development while improving the quality of life of the workforce and their families, the local community, and society in general. Core components of the CSR agenda are sustainable development, environmental protection, human rights, labour rights, equity, transparency, and genuine stakeholder engagement. CSR is not only about capacity building for preserving Arctic sustainable livelihoods but also promotes the balance of economic, social, and environmental aspects of corporate operations. It implies a sense of respect for cultural differences and adaptive ways of building skills of employees, empowering communities, and preserving the environment for future generations. CSR has a fundamental role to play in strengthening the relationship between businesses and communities in a sustainable way. The main topics of this chapter relate to CSR in the Arctic multicultural context, CSR and Arctic communities, energy, mining, agroecosystems, and gender and inequality.

Background: CSR in the Arctic multicultural context

The Arctic region is a sensitive environment composed of urban, rural, and remote communities. The crucial aspect of the debate on the development of responsible economic activities in a sensitive environment like the Arctic is reflecting on how to perform sustainable development in the Arctic. The challenge, nowadays, is to move towards a better-informed decision-making process based on data – documented and shared information on key economic sectors like agroecosystems, fishing, shipping, tourism, energy, mining, etc. The implementation of such a system would involve the identification of the main local and regional communities' needs, resources, capabilities, and sources of financing under a sustainable criterion. Scientific observational systems allied to Traditional Knowledge local systems would be recommended to respond to infrastructural but also to socio-environmental requirements. This community-based observation and knowledge building represents a step forward in the process of sharing values, as it is a coordinated effort to articulate scientific evidence and Traditional Knowledge to meet societal needs, increase resilience, reduce societal vulnerabilities, and promote climate change adaptation (Arruda and Krutkowski, 2017). Traditional Knowledge can be found in a wide variety of contexts including agricultural, ecological, medicinal, wildlife, ecosystems, and biodiversity-related knowledge (Arruda and Krutkowski, 2016).

CSR in the Arctic involves enhancing a framework of societal benefits (see Figure 6.1) relating to different areas.

It consists in adopting a consistent approach against the pressures posed by the challenges of these specific sectors and creating a robust framework of best practices based on the evaluation of societal impacts of specific case studies and the results of the efficient implementation of policies with practical economic, social, and environmental benefits. CSR is relevant to understanding and managing change based on different knowledge systems – scientific knowledge and Traditional Knowledge (Arruda, 2019).

CSR, extractive industry and Arctic communities: Energy systems, mining, agroecosystems

Energy systems

The Arctic presents a unique opportunity to innovate the global energy system and to evolve human adaptation based on a multi-level perspective (Geels, 2011). The challenges and risks posed by fossil fuel exploitation ventures and the additional layer of risks posed by climate change serves to amplify the understanding of how energy production/use is linked to local social processes and, ultimately, to sustainable development. The Arctic is a complex sensitive geographical area where geo–bio–physical components are entangled with socio-environmental factors. Mechanisms of adaptation which provide the right balance among economic, social, and environmental aspects are quintessential to this region's prosperity and durability.

This great potential for a sustainable prosperity project of co-management (Arruda, 2018a) was initially debated in *Renewable Energy for the Arctic: New*

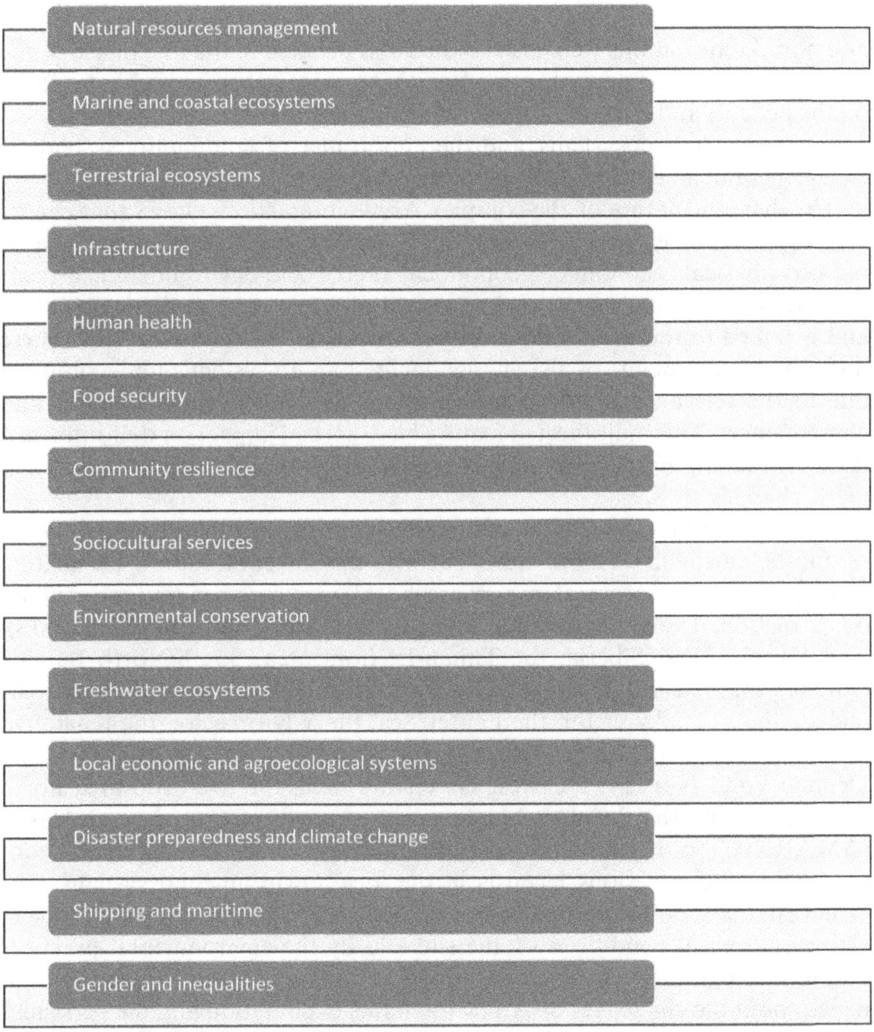

Figure 6.1 Societal benefits

Figure 6.1 explains the importance of CSR in the Arctic to enhance the framework of societal benefits that reflect the interaction of several components of different systems. This interaction is dynamic and the balance of the components have consequences for societal components.

Source: Based on Arruda and Krutkowski (2016)

Perspectives (Arruda, 2018b). In this book, the importance of a low-carbon Arctic energy system was widely discussed as well as the latest technological and innovative energy approaches from pioneering and leading stakeholders from different parts of the Arctic. Aspects like energy integration, infrastructure of communities, investments, corporations' projects in energy transition, energy policies and standards, and the main implications for Arctic and non-Arctic nations were also studied and debated in the first phase of the project 'Renewable Energy for the Arctic' (Arruda, 2018). This chapter focus on the benefits and effective impacts of the integration of smart grids on local communities and ecosystems, and the importance of community systematic engagement in an efficient renewable energy system.

The dynamic nature of the complex Arctic interactions shows that energy is a highly complex and interdisciplinary matter which involves a range of bio-geo-physical, economic, geopolitical, and socio-environmental aspects concomitantly. It has strong links to natural sciences as a physical process and is linked to the most vital biological processes in the food chain. These integrated components of the energy debate require systemic understanding due to the relevance of the complex interconnections of their causes and consequences. This multifaceted feature has a special impact on well-informed decision-making in the contemporary adaptation process accelerated by climatic changes.

In general, the Arctic region comprises vast amounts of extractive energy resources, mostly in offshore areas, but with the current trends of ice melting (cryosphere changes) several inshore areas are becoming open spaces for inland oil extraction. The main potential areas are the North Slope in Alaska (US); East Siberia, West Siberia, the Timan-Pechora area, South/North Barents (Russia); the area east of Norway and East Greenland. Other important areas include the Vilkitsky sector, the Laptev Sea, the Vilyuy sector, the Khatanga are, and the North Kara Sea (Russia); the West Barents Sea north of Russia; Denmark and Norway; the West Greenland area; the Sverdrup area north of Canada; and the Beaufort-Mackenzie area north of Canada and Alaska (AMAP, 2007, p. 10). Oil and gas exploration processes and accidental contamination create serious hazards in the local environmental systems. The extent of these polluting processes is influenced fundamentally by the special geographical conditions of the site and by the environmental practices and technologies used by the oil and gas companies. There is plenty of evidence about the challenges of Arctic fossil fuel exploitation and the persistent levels of pollution and toxicity caused by onshore and offshore hydrocarbon exploitation as well as by transboundary pollution from mid-latitudes that migrate to higher latitudes (AMAP, 2004, 2007, 2013, 2018; Arruda, 2014). A great concern resides in the fact that despite local and global regulation and standards, several sources of pollution and the extent of its impacts have not yet been completely assessed. The scale and complexity of changes in the Arctic have been revealed to be greater than the monitoring systems can cope with.

Apart from the mineral richness of this region, the scale and speed of the current regional transformation raises a sense of urgency (Overland et al., 2019). Sea-ice retreat and permafrost reduction have major implications for biogeochemical cycling, hydrology, vegetation, the food chain, human communities, and infrastructure. In figures from June and mid-July 2019, monitoring systems and satellite remote sensing showed that Arctic sea ice extent has been receding at record-low daily levels with the most significant ice losses in Hudson Bay and in the Kara Sea (SIPN, 2019). The general patterns of sea-ice extent in July 2019 were below the 1981–2010 average across the Arctic with just a few exceptions in the East Greenland Sea, around Svalbard, and within the East Siberian Sea, according to the latest data from the NASA Cryosphere website (Box et al., 2019). Specific weather systems data have shown a significant rise in temperatures in large extents of Alaska, which intensified impacts on regional ice-loss rates as well as triggering the latest unusual large wild fires due to the increase in temperature and precipitation correlated with increases in the number of lightning-caused wildfire ignitions (AMAP, 2019). Apart from these geophysical shifts, these extreme summer events impact both terrestrial and marine biological ecosystems (flora, fauna). These specific Arctic warming trends cause serious ecological impacts on plants' season dynamics as they flower earlier, affecting the survival of several other species dependent on degraded tundra ecosystems. Moreover, they are also affecting the demographics of animal species leading to increased mortality of wild reindeer populations among other species. The sea-ice loss has also triggered a significant impact on marine ecosystems as the shifts in the timing and intensity of marine algal blooms disrupts the life cycles of key marine species (krill, fish, birds, and mammals) responsible for the food web balance, including species such as polar cod, whales, seals, and polar bears (AMAP, 2017, 2019).

As Figure 6.2 shows, geo–bio–physical interactions are also linked to socio-cultural interactions and socio-environmental and economic impacts. The availability of many species that the Arctic indigenous people rely on for food (e.g. polar bears, seals, whales, and certain fish stocks) has become limited because of climate change and the receding ice cover. Additionally, warmer temperatures bring new bacteria and other microorganisms with unknown effects on human health or other species (Arruda and Krutkowsky, 2017). The lifestyle of North communities is significantly different from South communities and the high demands of the extractive industry affect not only subsistence but the cultural continuation of the northern peoples. These peoples need to be part of shaping their own adaptive approach to a changing Arctic, and energy systems are an essential part of this approach.

Successful developments of fossil fuel exploration and the mineral industry in the Arctic will probably alleviate the pressure and demands of southern markets and probably increase energy security. However, the continuation and expansion of the current extractive system in highly sensitive Arctic areas will drastically accelerate the degradation of ecosystems and intensify socio-environmental impacts of climate change locally and globally.

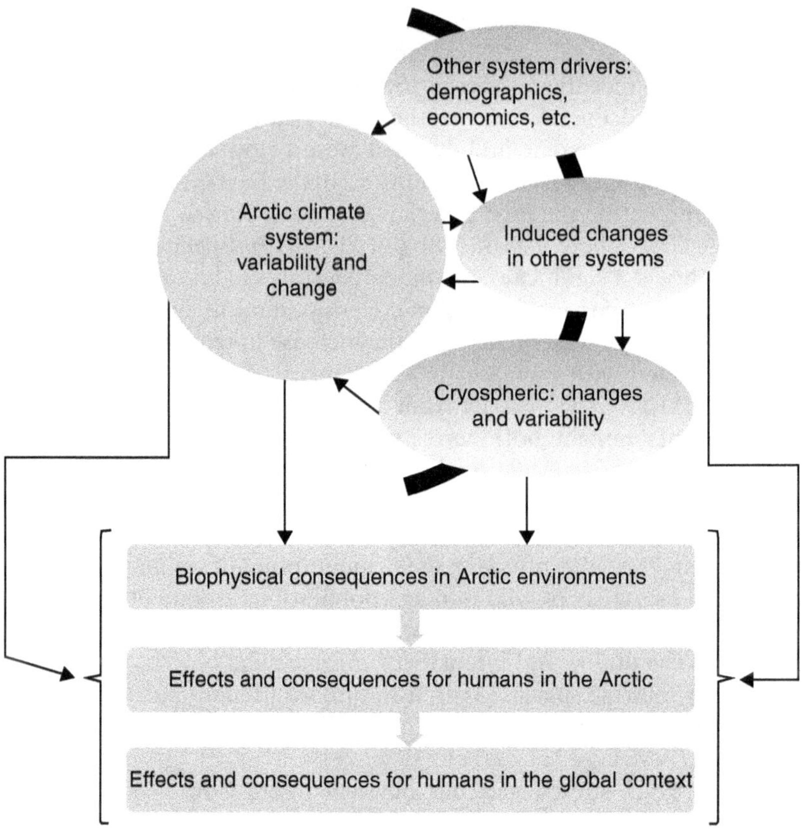

Figure 6.2 The Arctic cryosphere and linkages to various levels of effects or consequences of changes that compose the whole Arctic system

Figure 6.2 raises the important connection among several systems that so far have been treated in silos and in a fragmented way by science. These systems like the cryosphere, climatic system, societal and environmental systems interact in real time and imbalances in one imply imbalance and changes in the other systems. This figure is very relevant for explaining the point that systems are in dynamic equilibrium and need to be understood and studied within an integrated view.

Source: Adapted from Hovelsrud et al. (2011)

Energy transition and socio-technical systems

The challenges and risks posed by fossil fuel exploitation ventures and the additional layer of risks posed by climate change serve to enhance understanding of how energy production/use is linked to local social processes and, ultimately, to unsustainable patterns of development. The importance of a low-carbon Arctic energy system has motivated the debate and initiatives related to the latest technological and innovative energy approaches from pioneering and leading stakeholders from different parts of the Arctic. The research focuses on how energy integration, infrastructure of communities, investments, and corporations' projects in energy transition are shaping energy policies based on socio-technology theory and the well-being of Arctic communities. It has relevant implications for enhancing understanding of the impact of socio-technical systems in the Arctic.

This reflection is the outcome of eight years of fieldwork in the Arctic, which involved visiting urban and remote communities with the aim of understanding and mapping the Arctic energy systems and the opportunities and challenges of a renewable energy system for the Arctic. Insights from the existent literature were analysed jointly during interviews with scholars, community members, and energy professionals 'in loco', which helped to create a platform of ideas on the strategic paths of a renewable energy future for the Arctic region.

A broad perspective of Arctic renewable energy resources involves solar, wind, geothermal, hydro, and biomass, while the most popular renewable energy sources comprehend three basic types: solar energy (direct solar and wind), tidal energy, and geothermal energy. Obviously, the potential and modalities of energy developments vary on a regional scale, but this is the start of a systematic transformation based on transition management, community participation, investment, and regulation. Some Arctic nations have taken this invaluable opportunity to address the challenges of the current energy transition process. Countries like Norway, Iceland, the US (Alaska), Canada, and Greenland have been promoting effective changes in how the Arctic energy system is viewed by creating pilot projects representing a new energy platform based on renewable energy resources.

Several important modalities of renewable energy projects represent actual installed capacity across the Arctic. Geothermal developments in Icelandic geothermal reservoirs are the forerunner projects of geothermal energy in the Arctic (Hogenson, 2018). These projects provide consistent, affordable, and low environmental impact sources of electricity to the communities around them, but they are highly dependent on volcanic systems only present in specific geological areas in the world. River hydroelectric projects are also relevant, mainly in Norway which runs 99% of its electricity production on hydropower systems (Gebre et al., 2013). The great emphasis in terms of renewable energy projects however has been seen in rural Alaska remote electrical grid modernization focused on minimizing the use of diesel generators. The development of wind and solar systems with the immediate purpose of reducing the significant local energy costs of electricity generation and replacing inefficient and polluting diesel generators represent, on average, a reduction of more than 50 per cent in the costs of electricity; the poorest households spend up to 47 per cent of their income on energy, representing more than five times their urban neighbours (The Alaska Arctic Council Host Committee 2016, p. 14).

High costs, systemic inefficiency, and environmental impacts were factors that triggered a real shift in the Alaskan energy landscape. Based on the challenges faced by remote off-grid village populations, the Alaskan State legislature defined a new perspective through the Alaska Affordable Energy Strategy (AEA, 2016, p. 16). The aim of this strategy was to assess specific social energy needs of remote off-grid communities in order to enhance quality of life and provide autonomy by using available technology. Due to this strategy and co-related investments made by the Renewable Energy Fund (REF), wind and solar energy became a systematic part of the Alaskan energy matrix. Between 2008 and 2015, the Alaskan legislature provided $257 million to the REF. In 2018, $11 million from the Power Cost Equalization (PCE) fund was applied to finance eight new projects and, in 2012, REF was extended by the Alaskan legislature until 2022. In the 2019 figures, REF projects saved Alaskan communities more than $30 million in annual displaced fuel used for generation and heat. There are 63 REF grants in process and 42 already-funded projects under construction (AEA, 2019). A similar decarbonization agenda, but at a slower pace, can be seen in the Northwest Territories in Canada where 257 of the 292 remote off-grid communities are still reliant on diesel plants (Natural Resources Canada, 2011, 2017).

The serious implications of reliance on diesel for Alaskan and Canadian off-grid remote communities represent a great opportunity for adaptation of lifestyles to technical systems (Tushman and Anderson, 1986) and the discontinuation of 'business-as-usual' practices through a pattern of reorganization based on transition pathways and new technological trajectories (Nelson and Winter 1982; Geels and Schot, 2007) associated with specific social needs. The technological trends in energy are relevant to enable new pathways but, as Geels (2005a, p. 445) argues, "artefacts only fulfil functions in association with human agency and social structures". In other words, societal functions are met by socio-technical systems (Geels, 2002, 2005b, 2005c) defined by Geels (2004) as a cluster of elements like technology, regulation, practices, markets, cultural features, infrastructure, and networks, which are produced, reproduced, and refined by social actors. The perspective needs to be an integrated one.

The new perspective associated with the decentralization of electricity systems as a core component of Alaskan smart-grid energy projects illustrates how co-evolutionary processes of technology and societal issues are. The destabilization of Arctic systems created a window of opportunity for an incremental change where local actors decided to coordinate resources available to adapt to real-world pressures and needs (Smith et al., 2005) by designing transition paths and re-aligning to new social-technical systems. This transition path originated the so-called 'low-carbon Arctic smart community'. This is a new concept associated with the notion of a sustainable community in which its people have a strong sense of initiative and solidarity, with a solid capacity for mobilization and awareness of rights. Societies with these features also have high standards of community organization with a developed sense of self-management and sensitivity to cultural, social, economic, and environmental issues (Milbrath, 1989, p. 27). A sustainable community implies some principles like respecting and fostering the ethical principle of preserving all forms of life, preserving biodiversity, changing production and consumption patterns, reducing the use of non-renewable resources, encouraging the use of renewable resources, respecting the supporting capacity of ecosystems, and designing new integrated tools of environmental management.

The low-carbon Arctic smart community created from renewable energy systems illustrates the importance of realizing that sustainable development and a low-carbon society cannot be built up only based on technology. This is a narrow argument. Social, technological, economic, cultural, and managerial processes must co-evolve and are intrinsically connected in passive and active ways. A smart community based on smart grids can articulate systems comprised of diverse passive and active components involving various stakeholders, each of them performing specific functions through the incorporation of intelligent communication and management systems (Giordano et al., 2011).

Consequently, policymakers will face a real challenge of shifting the perspective of the current passive energy system (supply-driven energy mode) to a multi-level energy perspective where the energy system is integrated enough to allow the active performance of different stakeholders contributing to the overall functioning of the energy system as a socio-economic unit. This active stakeholder performance means that the community (citizens, businesses, government, entrepreneurs, suppliers, etc.) understands the meaning of energy systems, energy production, use and demand, energy targets, and the relevance of renewable energy sources in the context of a low-carbon economy for industrial, agricultural, public sector, commercial, and household electricity and heat users. They understand the concept and implications of energy efficiency, eco-design, low-carbon transport systems, and water and heating-saving programmes, and they also know the mechanisms to perform and achieve the established targets (local, regional, international), taking an integrated view of other systems dependent on energy sources. They are active parts in living a low-carbon lifestyle compatible with a low-carbon economy based on green jobs, education, and technical training to operate in a low-carbon integrated system. This perspective involves a renewable energy leadership oriented to more well-defined paths for a global low-carbon society and a low-carbon economic system.

The new perspective resides on the fact that one cannot only look at this renewable energy revolution from the viewpoint of technological progress. Instead, the great potential of a sustainable co-management prosperity project involves the co-evolution of technology and society (Geels, 2005b) within a multi-level perspective (Geels and Shot, 2007) of interconnected interactions and functionalities. This co-evolution involves interfaces between technology, users, market, science, institutions, and cultural and environmental factors. In other words, technology does not represent a solution by itself but is a tool to approach the complexity and dynamicity of systemic interactions in practice. The socio-technical systems in renewable energy represent an active and holistic integrative process that allows low-carbon smart communities to exist and prosper while respecting the support capacity of the planet.

Energy transition is also a transition of socio-technical systems (Arruda and Arruda, 2018). The technological and innovative energy approaches from the pioneering Alaskan Renewable Energy smart grids not only have shifted energy production and use from fossil fuels to a more sustainable path, but also have provided a new horizon of durable energy access to Alaskan remote communities. The Alaskan smart-grid projects reflect a great effort of local expertise and make a significant contribution to managing local and regional energy demand flows, optimizing energy use, minimizing fossil fuel impacts from diesel generators, and creating a low-carbon Arctic smart community based on a decentralized type of renewable energy system relying on advanced storage and metering systems. The stabilization, efficiency, and integration of the Alaskan Renewable Energy smart grids offer new challenges and opportunities for the years ahead. It is the ultimate case study of socio-technical energy systems co-evolving with smart technological solutions, co-management approaches, and contemporary energy policies with a perspective of application at a global scale.

Global application and emerging cases of socio-technical systems

Envisioning a sustainable society is important but it is presently not enough considering the latest scientific evidence and the scale of climate change impacts. There is global concern about the causes and consequences of climate change and persistent intangible rhetoric about sustainable development while different global actors seem to take distinct paths in relation to these contemporary dilemmas. As remarked in the previous section, the Alaskan Renewable Energy smart grids reflect an important trend. Local experts from different sectors of society have committed strongly to a low-carbon economy on a practical basis, and the smart grids represent a significant contribution to managing local and regional challenges. It has been proved possible to articulate stakeholders' responsibility and adopt proactive initiatives to decrease fossil fuels use and greenhouse gases emissions based on renewable energy resources. It also demonstrated the importance of community engagement and understanding of the process to effectively contribute to energy-efficiency programmes and measurable change in production and consumption patterns.

Other areas of the world are pursuing the same path despite showing different outcomes. An emerging similar case to the Alaskan low-carbon communities is Ireland's 'sustainable energy communities', a result of the vanguard national policy (DCENR, 2015) and national mitigation plan. Based on proposed mitigation commitments of the European Union (EU) target of reducing at least 40% of its domestic GHG emissions by 2030, the Sustainable Energy Authority of Ireland (SEAI) has influenced local energy production and use by advocating the vision of transforming Ireland into a society based on sustainable energy structures, technology, and practices (SEAI, 2016).

Based on the National Mitigation Plan of 2017 (DCCAE, 2017a), Ireland's position is fundamentally focused on engaging, equipping, and integrating communities to meet energy targets and leading the country into a low-carbon future (SEAI, 2015; Sampath et al., 2018) based on socio-technical transition, community engagement and impact. Targets achieved account for more than 6 million tonnes of CO_2 emissions avoided every year since the beginning of the plan in 2015 through energy efficiency (DCCAE, 2017b) and renewable energy technologies and programmes (SEAI, 2017). Over 350,000 homes, 3,500 businesses and public-sector entities have saved approximately €1 billion annually based on energy efficiency measures supported by local government programmes. Although significant progress has been made, there are still several challenges to overcome through energy education, multi-sectoral infrastructure adaptation and integration, and the creation of green jobs. However, this is the start of a trend that goes beyond mere technological achievements to configure a genuine societal shift, confirming Geels' (2004) vision of a sustainable energy system based on technology, regulation, practices, markets, cultural features, infrastructure, and networks, refined by social actors.

Mining in the Arctic

As the Arctic ice recedes, new opportunities for mineral extraction and shipping routes have emerged. Nations, even those not in the Arctic themselves, are claiming their portions of this wealth to supply the world's high demand for resource such as rare earths, gold, phosphate, bauxite, iron ore, copper, nickel, diamond, and oil. Countries leading specific industries such as batteries, mobile phones, computers, and the military are all claiming a share of the Arctic's abundant natural resources. This raises geopolitical tensions as well as a race for natural resources. The outcomes of this new mineral race bring all sorts of concerns not only for the locals but also to the world. Apart from important locations in the Arctic that have explored mining activities over the last century or more, new locations have emerged, i.e. Greenland. The government of Greenland, meanwhile, has expressed interest in economic diversification and in attracting foreign investment to expand its mining infrastructure projects to increase regional revenue as well as become less reliant on the fishing industry.

Despite the strategic environmental and social risk assessments (SEA/EIA and SSIAs) prepared by energy and mining companies towards gaining licences to exploit resource-rich areas (Arruda and Krutkowski, 2016, 2017) and despite the specific regulations regarding transboundary pollution and environmental toxicity according to the best international standards and practices (NORSOK – Norwegian Safety Regulations; Environmental Benefit Analysis; NORSOK, 2010), the significant increase in mining activity in Greenland will require a systematic Greenlandic sustainability policy in place to guide industries and planners. This is an example of economic activity that requires a clear CSR strategy to protect the environment and Arctic society. Ideally, the Greenlandic government in partnership with its main trading nations (Denmark, the EU, the US, China) should work together to ensure the rigorous delivery of the United Nation's framework of Sustainable Development Goals (SDGs) (United Nations General Assembly, 2015).

Agroecosystems and food security

The rich food chain in the Arctic has sustained local communities for hundreds of years. Based on traditional practices, Arctic communities have been living for generations according to the principle of not taking more than needed from the environment (Arruda, 2019). A changing climate may alter the environment for traditional patterns of hunting and gathering, while permafrost dynamics, native vegetation, and soil temperatures also create the potential for new crops in different areas of the circumpolar region. Climate change may also alter the diet of many indigenous Arctic peoples as groups used to eating meat are consuming more grains and vegetables, increasing demand for locally grown foods. These changes have the potential to unlock the local northern agriculture, but it will depend on new economic models and food production programmes, and northern agricultural science, practices, and policies.

In the Arctic, warmer weather and frost-free seasons have allowed enterprising agricultural researchers to grow a variety of vegetables, grains, herbs, and other plants typically planted in temperate fields. Experimental farms can be seen in Alaska, Canada, Greenland, and Norway in places where agriculture was not possible before. Livestock production (sheep) is also favoured with the production of hay and animal grass forage.

Contemporary local food production seems an auspicious idea, taking into consideration that in some Arctic areas there are remote communities not served by roads and dependent on a carbon-intensive network of transportation modes to bring food and supplies to communities. The price of these foods is disproportionally high because they are shipped long distances and subject to natural and human-made disasters (i.e. earthquakes, volcanoes, storms, and shipping strikes). These supply chains can be disrupted and isolated communities cut off, which reinforces the idea of a more unified approach to local food production.

The idea of sustainable food systems in the Arctic emphasizes the concept of circularity and resource efficiency of food systems for providing fresh fruit and vegetables instead of carbon-intensive frozen food sealed in plastic. It is about nutrition for healthier diets but also climate-smart and environmentally sustainable food systems. As an important effect of climate change, the local economies and diets of many indigenous Arctic peoples are also changing because groups that used to consume meat are now consuming more grains and vegetables, increasing the demand for locally produced foods (Nobel, 2013), such as tomatoes, leeks, and barley being produced in Alaska, Greenland, and Norway. According to Torfi Jóhannesson, president of the Circumpolar

Agricultural Association, "agriculture will be possible in many places where it is not now" due to the longer and warmer growing season (Hoag, 2016). This sector offers an invaluable opportunity for entrepreneurs and companies operating in the Arctic to design CSR projects involving smallholders and farmers by using the local potential (geothermal greenhouses) to develop more sustainable and greener local economies and food chains.

Livestock production is also moving to the North with a prospect of expansion of more than 50% in the next 50 years, but there are also challenges concerning pests and diseases moving into northern niches, climatic events like flooding in farmlands, as well as mining sites and their radioactive and chemical waste that can contaminate soils and water resources. This is particularly sensitive in south Greenland, where there is an important debate about jobs, revenue, and local economy with anti-uranium groups arguing the importance of stimulating agricultural activities instead of mining extraction.

Strategic partnerships between governments, business, and universities would be important sources of investment, technological support, and know-how (knowledge creation/dissemination) to advance research and development to identify productive cultivars, conduct soil surveys, and promote local adaptation by selecting the most adequate sites for agriculture under specific social-environmental risk assessments. These actions, therefore, require a meaningful engagement of stakeholders (OECD, 2016) and further investigation about the quality of decision-making in CSR and local perceptions of CSR.

The idea of sustainable food systems in the Arctic relates to the concept of circularity and resource efficiency of food systems but also how to manage change by sustainably unlocking the potential and expansion of northern agriculture, reducing the negative effects of international trade to the Arctic and managing cultural diversity in terms of food production.

Inequality as a result of drivers of change

In general, the Arctic has always been associated with males because of the ancient explorers, hunters, and reindeer farmers. In the literature, most part of the texts about the Arctic reflect men's views about exploration, military activities, and discoveries in a harsh environment. The history and literature are permeated by westernized perspectives and colonial regimes' male privileges.

Even nowadays, current discourses reflect a very masculine agenda of economic exploration, military security, and modernization. This historical dynamic and the environmental changes in the region have triggered significant societal transformation that deserve further research. The most significant societal changes can be identified as the changing roles of men and women in terms of leadership, representativeness, resource management, and regional development. The gender perspective in Greenlandic society seems to be an additional growing topic of debate strictly interconnected with development perspectives and level of education. In other words, gender balance or equality need to be part of a detailed reflection about the circumpolar development model and education.

History has shown that gender inequality has been a common pattern in different cultures with no exception to indigenous communities due to the diversity of their cultural practices or because of their interaction with the majority society. Peoples have a collective right to self-determination and culture, which signifies that they are free to choose their political, economic, and cultural organization models. Collective rights must be interpreted in a way that generally promote the individual rights of the members of the group (Stavenhagen, 2002). Many of the challenges in the Arctic are of a global and developmental nature.

In terms of rights and legal framework, the ILO Convention n. 169 is iconic (ILO, 1989). Launched in 1989 by the International Labour Organization (ILO), the 'Indigenous and Tribal Peoples Convention' has been ratified in more than 20 countries and seemed particularly relevant by triggering the process of multicultural understanding and consequent protection of indigenous people's rights based on case studies and key historical experiences. ILO n. 169 does not define indigenous and tribal people but identifies key characteristics, providing objective criteria to determine the protected groups. These characteristics or elements of Indigenous Peoples include historical continuity since precolonial societies, ancient territorial attachment, language, way of living, culture, and self-identification.

Another relevant document in relation to indigenous protection is the UN Declaration on the Rights of Indigenous Peoples (UNDRIP, 2007), which does not include a definition but has identifying characteristics attributed to Indigenous Peoples and their relationship with lands, natural resources, territories, language, and traditions.

Despite the existence of these and other relevant documents comprising a legal and political framework to identify and protect indigenous communities, there are many constraints and difficulties in collecting and interpreting data regarding the main societal changes affecting indigenous communities. Another tendency is the increasing number of indigenous migrants moving to urban centres, assuming new roles in their specific human groups and manifesting new expressions of identity contributing to the transformation of community structures.

The first concrete document that launched the gender debate was the first Arctic Human Development Report published in 2004 (AHDR, 2004; Arctic Council, 2017). The chapters of this inaugural document started to address important aspects of the critical gender imbalance reflecting major societal changes in the North. Despite identifying key issues like migration, gendered violence, lack of political representation, and educational and economic negative contexts, no further progress has been made since then.

In general, international treaties seem to have a positive impact but Nordic policies and legislation have not been explicit about gender equality and even constitutional texts presenting equality provisions have not been effectively implemented to prevent gender discrimination. The major part of these legal texts reflects colonial provisions stressing the disadvantageous position of women and ethnic minorities. Consequently, it is possible to state that this economic and political status is embedded in general and specific legislation, demonstrating that it is an enrooted social phenomenon in need of a social engaging transformative strategy.

Considering the potential benefits and challenges offered by resource extraction, transportation, energy development, and infrastructure in the North, these transformative strategies tend to pass by the design and application of effective resource management models more focused on improving quality of life, income distribution, and the stability of communities. Because of human capital investment, societies could achieve wealth creation and better levels of equality.

In places where economic mechanisms are used to boost revenues in a rapid way, without being used to enhance social capital and human development, economic imbalances tend to create and worsen gender and ethnical inequalities, marginalization, and poverty.

Most of the economic activities being developed in Greenlandic communities show the pattern of male dominance. Industries like hydrocarbon (oil and gas) exploitation, mining, and infrastructure currently generate 77% of employment opportunities for men, reallocating a great parcel of social capital previously dedicated to traditional activities to the new economic hubs (Kleist et al., 2015).

The neoliberal governance style seems to stimulate economic and social asymmetries as emphasis has been given to the exploitation of non-renewable resources (hydrocarbons and minerals) that provide a temporary source of royalties paid to local governments and open space for developers to operate on indigenous lands, but does not present a good strategy for the impermanent economic reality caused by resource exhaustion. It means that job opportunities, revenue flow, and environmental decisions are not in the end sustainable.

This unsustainability has induced underdeveloped societies left to live with serious social and economic externalities because of the development model practised. Gender and ethnic inequality are a consequence of the imbalanced model of development that only invests in short-term economic gain but does not privilege the social and environmental pillars of sustainability (Elkington, 1994, p. 92).

Development models have changed not only the environment but also the social tissue. The difference is that western science has given more emphasis, over the last century, to economic and most recently to environmental patterns but have not afforded the same attention to social transformations.

Modernization brought on board diverse lifestyle aspirations, different local dynamics related to environmental change, new economic activities, and subsistence strategies. Over a few generations, the range of changes in Arctic societies have shown a direct impact on family standards, sizes, and household and gender roles. Originally based on traditional subsistence, Greenlandic society currently envisages self-employment as a less viable livelihood model.

The first decisive cause of impact was the introduction and development of the welfare state in Greenland, determining the Greenlandic family structure as similar to the westernized family model. The structure has to do with the size, gender roles, employment and levels of education, contribution to the household, and general aspirations. According to Poppel (2015), the distribution of different tasks, duties, and rights, as well as expectations related to gender have changed. A consequence of these changes is that many women have had better possibilities to get an education and to participate in highly paid labour market activities, while the occupational structure related to men has shown a decline in the share of hunting and fishing activity but an increase in male participation in construction and retail trade.

The development process with its positive and negative impacts triggered significant social changes in the occupational structure dominated by men's activities, identified primarily as hunting and fishing and, currently, absorbed by the labour market with a central aspiration of contributing to the family's income for paying the bills.

In Greenland, gender became a crucial factor to be studied. Gender is emerging as a central social parameter to be addressed by stakeholders and decision-makers in relation to Greenland's current development model. Due to the diversity of issues associated with the gender phenomena, several communities are being left behind in terms of socio-economic development because of educational gaps, migration from the North to the South, and employment opportunities.

Over the last two decades, new industries and business models started to be developed in Greenland despite the small-scale Greenlandic economy. In 2011, approximately 3,860 companies were based in Greenland. 75% are owner-operator firms with fishing being the main industry, generating 93% of exports (Kleemann, 2018). A significant number of other companies are fully or partially owned by the government, including infrastructure, energy, and administration employing approximately 6,000 people.

Occupation by gender in Greenland shows a segregated labour market with a wide participation of women in politics and there are several cases of single mothers actively working as the main income providers in a new family structure. Currently, there are other social issues contributing to the continuous change in the social structure and gender roles in Greenland, such as migration, housing shortages, and low-skill workers with no chance of entering the labour market. They also lack the specific opportunities to acquire necessary skills for an even more sophisticated labour market of energy and mineral resources exploitation involving highly complex projects and operations requiring specific knowledge of health and safety measures.

In designing strategies to balance gender inequality derived from economic development, it is important to consider the relationships between economic growth and human development. Women and men do not necessarily benefit

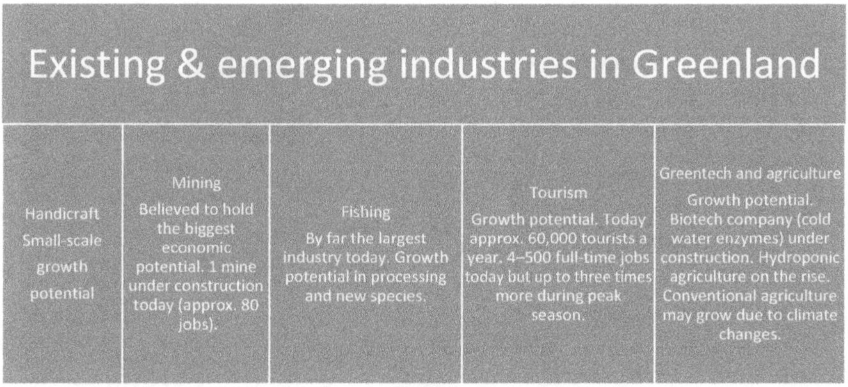

Figure 6.3 Existing and emerging industries in Greenland

Figure 6.3 briefly presents the most important productive activities in Greenland representing a source of significant revenue to the Greenlandic economy. The blend of traditional and emerging economic activities also represents a source of impacts on the local environment and on society. The figure also projects future economic prospective activities.

Source: Kleist et al. (2015)

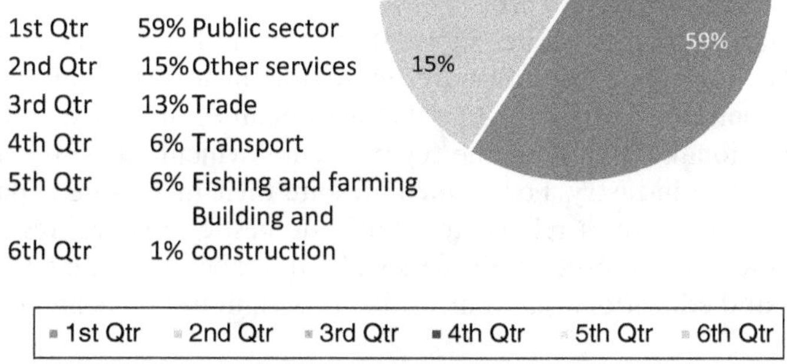

Figure 6.4 Women's employment by industry in Greenland

Figure 6.4 presents data collected in the field and illustrates the proportion of activities performed by women in Greenland. It shows the share of economic activities and the respective percentage of women in sectors like trade, fishing, construction, public sector, transport, and farming. This is an important figure showing data that illustrates the arguments related to gender and diversity in Greenland.

Source: Kleist et al. (2015)

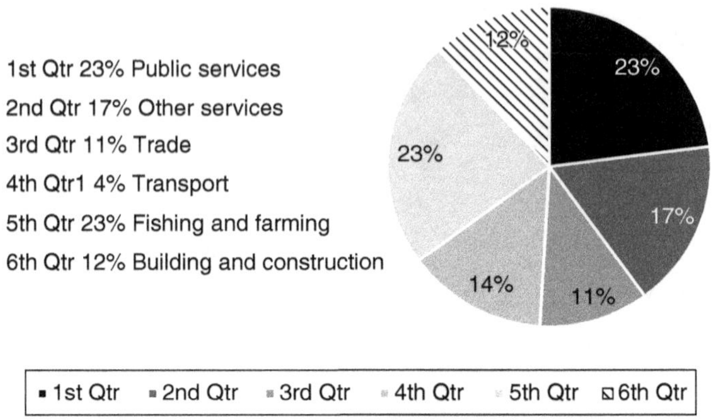

Figure 6.5 Men's employment by industry in Greenland

Figure 6.5 presents data collected in the field and illustrates the proportion of activities performed by men in Greenland. It shows the share of economic activities and the respective percentage of men in sectors like trade, fishing, construction, public sector, transport, and farming. This is an important figure showing data that illustrates the arguments related to gender and diversity in Greenland.

Source: Kleist et al. (2015)

from such economic growth and are affected unequally by such development. In addition, the traditional economy and women's and men's participation in this economy must be accounted for in the development of both short-term and long-term sustainable economic policies. The question is if the speed of societal changes and the demand for natural resources will allow enough time for societies to understand, map, and adapt to their geographical changes in time to promote the desired balance.

A gender analysis of these issues shows that women play significant roles in Inuit society; however, these roles are not visible in formal institutional structures. For men, the analysis highlights the difficulties of making a transition from traditional activities to the types of employment currently available in the extractive industry. For women, despite their increasing capacity of adaptation, serious issues related to health, domestic violence, resource management, and better representativeness will require better-designed strategies (political and educational) and investment to minimize risks and create opportunities.

The active participation in the design of an innovative and long-term development process seems to be a clear path to address inequalities created by the drivers of change. Knowledge-based jobs, in both public and private sectors, will help the transition to a modern Greenlandic society and a long-term development model. Entrepreneurship (multicultural enterprises) can be a vital trend in this transition through small local knowledge-based companies that may be able to provide new possibilities for a diverse job market by attracting both women and men with highly relevant competences to build up Greenland's future (OECD, 2011). Capacity building and education oriented to sustainability are key components to start designing the future of a more balanced Greenlandic development model. It is a time-demanding task that will require planning and expertise.

Questions for discussion

1. What does it mean to develop a low-carbon Arctic smart community?
2. What does it take to promote CSR in the Arctic multicultural context?
3. Why is it correct to say that energy transition is also a transition of socio-technical systems?
4. How can CSR tackle local inequalities caused by drivers of change?

Chapter summary

The Arctic region, as a sensitive environment composed by urban, rural, and remote communities inspire a crucial debate on the patterns of development of responsible economic activities that can perform sustainable development in the Arctic. The active participation in the design of an innovative and long-term development process seems to be a clear path to address inequalities created by the drivers of change. The implementation of a responsible development system would involve the identification of the main local and regional communities' needs, resources, capabilities, and sources of financing under a sustainable criterion based on strategies to address socio- technical components, balance inequalities and apply effective resource management models focused on improving quality of life, income distribution and the stability of communities.

References

Alaska Arctic Council Host Committee (2016). *Environmentally Responsible Resource Use and Development in the U.S. Arctic*. Alaska: Author.

Alaska Energy Authority (AEA). (2016). *Annual Report*. Accessed on 1 July 2019 from www.akenergyauthority.org/Portals/0/Publications/AEA2016AnnualReport.pdf?ver=2017-03-01-180130-100

Alaska Energy Authority (AEA). (2019). *Renewable Energy Status Report*. Alaska, US: Author.

Arctic Council. (2017). Fairbanks Declaration, Arctic Council Secretariat. Accessed on 10 July 2019 from https://oaarchive.arctic-council.org/handle/11374/1910

Arctic Human Development Report (AHDR). (2004). Akureyri, Iceland: Stefansson Arctic Institute.

Arctic Monitoring and Assessment Programme (AMAP). (2004). *Impacts of a Warming Arctic*. Cambridge, UK: Arctic Climate Impact Assessment (ACIA).

Arctic Monitoring and Assessment Programme (AMAP). (2007). *Arctic Oil and Gas 2007*. Oslo, Norway: Author.

Arctic Monitoring and Assessment Programme (AMAP). (2013). *AMAP Assessment 2013: Arctic Ocean Acidification*. Oslo, Norway: Author.

Arctic Monitoring and Assessment Programme (AMAP). (2017). *Snow, Water, Ice and Permafrost in the Arctic (SWIPA) 2017*. Oslo, Norway: Author.

Arctic Monitoring and Assessment Programme (AMAP). (2018). *Biological Effects of Contamination on Arctic Wildlife & Fish Summary for Policymakers*. Tromso, Norway: Author.

Arctic Monitoring and Assessment Programme (AMAP). (2019). *Arctic Climate Change Update 2019*. Tromso, Norway: Author.

Arruda, G.M. (2014). Global governance, health systems and oil and gas exploration. *International Journal of Law and Management*, 56(6), 495–508. https://doi.org/10.1108/IJLMA-06-2014-0040

Arruda, G.M. (2015). Arctic governance regime: the last frontier for hydrocarbons exploitation. *International Journal of Law and Management*, 57(5), 498–521.

Arruda, G.M. (2018a). Arctic resource development: A sustainable prosperity project of co-management. In G.M. Arruda (Ed.), *Renewable Energy for the Arctic: New Perspectives* (pp. 106–117). Abingdon, UK: Routledge.

Arruda, G.M. (2018b). *Renewable Energy for the Arctic: New Perspectives*. Abingdon, UK: Routledge.

Arruda, G.M. (2019) *Sustainable Energy Education in the Arctic: The Role of Higher Education* (1st ed.). Abingdon, UK: Routledge.

Arruda, G.M. and Krutkowski, S. (2016). Arctic governance, indigenous knowledge, science and technology in times of climate change: Self-realization, recognition, representativeness. *Journal of Enterprising Communities: People and Places in the Global Economy*, 11(4), 514–528.

Arruda, G.M. and Krutkowski, S. (2017). Social impacts of climate change and resource development in the Arctic: Implications for Arctic governance. *Journal of Enterprising Communities: People and Places in the Global Economy*, 11(2), 277–288. https://doi.org/10.1108/ JEC-08-2015-0040

Arruda, G.M. and Arruda, F.M. (2018). Towards sustainable energy systems through smart grids in the Arctic. In G.M. Arruda (Ed.), *Renewable Energy for the Arctic: New Perspectives* (pp. 87–105). Abingdon, UK: Routledge.

Box, J. et al. (2019). Key indicators of Arctic climate change: 1971–2017. *Environmental Research Letters*. https://doi.org/10.1088/1748-9326/aafc1b

Bourguignon, D. (2016). *Closing the Loop. New Circular Economy Package*. European Parliamentary Research Service. Briefing Jan. 2016, Members' Research Service.

Department of Communications, Climate Action and Environment (DCCAE) (2017a) *National Mitigation Plan*. July 2017, Ireland.

Department of Communications, Climate Action and Environment (DCCAE) (2017b) *National Energy Efficiency Action Plan 4, 2017–2020*. Accessed on 12 July 2019 from www.dccae.gov.ie/documents/NEEAP%204.pdf

Department of Communications, Energy and Natural Resources (DCENR) (2015). *Ireland's Transition to a Low Carbon Energy Future*. (Department of Communications, Energy and Natural Resources – now DCCAE). Accessed on 1 July 2019 from www.dccae.gov.ie/documents/Energy%20 White%20Paper%20-%20Dec%202015.pdf

Elkington, J. (1994). Towards the sustainable corporation: Win-win-win business strategies for sustainable development. *California Management Review*, 36(2), 90–100.

Gebre, S., Alfredsen, K., Lia, L., Stickler, M., and Tesaker, E. (2013). Review of ice effects on hydropower systems. *Journal of Cold Regions Engineering*, 27(4), 196–222.

Geels, F.W. (2002). Technological transitions as evolutionary reconfiguration processes: A multi-level perspective and a case study. *Research Policy*, 31, 1257–1274.

Geels, F.W. (2004). From sectoral systems of innovation to socio-technical systems: Insights about dynamics and change from sociology and institutional theory. *Research Policy*, 33 (6–7), 897–920.

Geels, F.W. (2005a). The dynamics of transitions in socio-technical systems: A multi-level analysis of the transition pathway from horse-drawn carriages to automobiles (1860–1930). *Technology Analysis & Strategic Management*, 17(4), 445–476.

Geels, F.W. (2005b). Co-evolution of technology and society: The transition in water supply and personal hygiene in the Netherlands (1850–1930) – a case study in multi-level perspective. *Technology in Society*, 27, 363–397.

Geels, F.W. (2005c). *Technological Transitions and System Innovations: A Co-evolutionary and Socio-Technical Analysis*. Cheltenham, UK: Edward Elgar.

Geels, F.W. (2011). The multi-level perspective on sustainability transitions: Responses to seven criticisms. *Environmental Innovation and Societal Transitions*, 1, 24–40.

Geels, F.W. and Schot, J. (2007). Typology of sociotechnical transition pathways. *Research Policy*, 36(3), 399–417.

Giordano, V., Gangale, F., and Fulli, G. (2011). *Smart Grids Projects in Europe: Lessons Learned and Current Development*. JRC Reference Report, EUR 24856EN. Petten, the Netherlands: European Commission, JRC.

Hoag, H. (2016). Arctic agriculture: Farming opportunities on the horizon. Accessed on 23 May 2020 from www.newsdeeply.com/arctic/articles/2016/10/31/arctic-agriculture-farming-opportunities-on-the-horizon.amp

Hogenson, J.M. (2018). Geothermal energy development in the Arctic. In G.M. Arruda (Ed.), *Renewable Energy for the Arctic: New Perspectives* (pp. 43–67). Abingdon, UK: Routledge.

Hovelsrud, G.K., Poppel, B., Oort, B., and Reist, J.D. (2011). Arctic societies, cultures, and peoples in a changing cryosphere. *AMBIO*, 40(1), 100–110.

Kleemann, N. (2018). *Greenland in Figures 2018*. Statistics Greenland (15th revised ed.). Greenland: Statistics Greenland.

Kleist, K., Jacobsen, M., Nielsen, P., Jacobsen, M.T., and Knudsen, R.J. (2015). *Everybody on Board. The Human dimension A Greenland Perspective on Capacity Building in the Arctic*. Greenland: A Greenland Perspective publication.

Milbrath, L. (1989). *Envisioning a Sustainable Society: Learning Our Way Out*. Albany, NY: State University of New York Press.

NASA. (2019). NASA Cryosphere website. Accessed on 20 July 2019 from https://neptune.gsfc.nasa.gov/csb/index.php?section=54

Natural Resources Canada (NRCan). (2011). Status of Remote/Off-Grid Communities in Canada. Accessed on 20 June 2019 from www.nrcan.gc.ca/energy/oublications/sciences-technology/renewable/smart-grids/11916

Natural Resources Canada (NRCan). (2017). Comprehensive energy use database. Accessed on 20 June 2019 from http://oee.nrcan.gc.ca/corporate/statistics/neud/dpa/menus/trends/comprehensive_tables/list.cfm

Nelson, R.R. and Winter, S.G. (1982). *An Evolutionary Theory of Economic Change*. Cambridge, MA: Belknap Press.

Nobel, J. (2013). Farming in the Arctic: It can be done. Accessed on 12 December 2020 from https://modernfarmer.com/2013/10/arctic-farming/

NORSOK (2010). *Z-013 Risk and Emergence Preparedness Assessment*. Lysaker, Norway: Author.

Organisation for Economic Co-operation and Development (OECD). (2011). *Guidelines on Multinational Enterprises* (2011 edition). Paris: Organisation for Economic Co-operation and Development Publishing. Accessed on 12 November 2018 from www.oecd.org/corporate/mne/

Organisation for Economic Co-operation and Development (OECD). (2016). *Due Diligence Guidance for Meaningful Stakeholder Engagement in the Extractives Sector*. Paris: OECD Publishing.

Overland, J. et. al. (2019). The urgency of Arctic change. *Polar Science*, ISAR5 Special Collection. https://doi.org/10.1016/j.polar.2018.11.008

Sampath, A., Lohan, J., and Costello, G. (2018). Assessing the role of energy intermediaries in Ireland's first generation of sustainable energy communities, *Proceedings of the 13th Conference on Sustainable Development of Energy, Water and Environment Systems*, SDEWES2018.0195, 1–11.

Sea Ice Prediction Network (SIPN). (2019). *The Sea Ice Outlook 2019/July Report*. Washington, D.C. Accessed on 20 July 2019 from www.arcus.org/sipn/sea-ice-outlook/2019/july

Smith, A., Stirling, A., and Berkhout, F. (2005). The governance of sustainable sociotechnical transitions. *Research Policy*, 34, 1491–1510.

Stavenhagen, R. (2002) Indigenous peoples and the state in Latin America: An ongoing debate. In R. Sieder (Ed.), *Multiculturalism in Latin America*. Institute of Latin American Studies Series. London: Palgrave Macmillan. https://doi.org/10.1057/9781403937827_2

Sustainable Energy Authority of Ireland (SEAI). (2015). *Unlocking the Energy Efficiency Opportunity*. Accessed on 13 July 2019 from www.seai.ie/resources/publications/Unlockingthe-Energy-Efficiency-Opportunity-Summary-for-Policymakers.pdf

Sustainable Energy Authority of Ireland (SEAI). (2016). *Ireland's Energy Targets: Progress, Ambition & Impacts*. Dublin, Ireland: Author.

Sustainable Energy Authority of Ireland (SEAI). (2017). *Ireland's Energy Projections: Progress to Targets, Challenges and Impacts*. Dublin, Ireland: Author.

Tushman, M. and Anderson, P. (1986) Technological discontinuities and organization environments. *Administrative Science Quarterly*, 31, 465–493.

United Nations Declaration on the Rights of Indigenous Peoples (UNDRIP). (2007). United Nations Declaration on the Rights of Indigenous Peoples. United Nations Permanent Forum on Indigenous Issues. Accessed on 25 January 2021 from www.un.org/development/desa/indigenouspeoples/wp-content/uploads/sites/19/2018/11/UNDRIP_E_web.pdf

Wilson, E. et al. (2016). *Meaningful Community Engagement in the Extractive Industries*. London: IIED.

7 CSR and practical solutions in complex environmental contexts

Learning objectives

- Understand the relationship between the environment and sustainability.
- Understand the intrinsic connection between the environmental and social dimensions.
- Understand the relevance of ethics and emerging technologies, including Responsible Research and Innovation.
- Understand how technology can benefit the environment and communities.
- Understand systemic thinking and resilience thinking.
- Understand the concept and implications of the circular economy.

Introduction

In the Earth's system, nothing lives in isolation. The processes are interconnected and any intervention in the South affects the North and vice versa. Problems and solutions are also part of the same system and are intrinsically interconnected. In this sense, it is crucial to understand that a system is a set of interconnected entities that are part of an integrated whole, and the key element of 'systems thinking' is considering that a system has boundaries for which it is crucial to define rules to safely operate the system. This is the main concern in globalized economies and societies in which a deteriorating global environment seems to have reached a point of no return, posing significant threats to environmentally sustainable development.

With an exponentially growing population with all sorts of needs, demands, and waste production associated with its large-scale industrial extraction and production, the pressures on the environment are growing proportionally and continuously. Our modern global society is facing an unprecedented escalation in global demand for energy, food, water, and other natural resources. At the same time, insecurity of supply sources and continued dependence on fossil fuels for energy generation and transportation is also escalating. These are factors are responsible for depleting natural resources and a steady increase of carbon dioxide emissions, which experts like Helm (2007, pp. 20–21) and

Stern (2007, pp. 20, 65, 68) believe is responsible for increasing average global temperatures and triggering cyclical climatic variations.

Environmental challenges are complex and interlinked and can trigger a chain of additional impacts on social and economic systems locally and globally because economic growth and enhanced societal well-being depend on ecological factors. At a time in the 21st century when science points out that several systemic boundaries have been breached (Steffen et al., 2015) and there is a predominance of environmental pressures over other global threats, we seem to have reached the turning point in which resilience strategies will need to be designed and implemented in the short, medium, and long term.

Types of discussion and issues explored in Chapter 7

This chapter will focus on the most pressing environmental issues of our times and the intrinsic connection among environmental, social, and economic systems. The discussion will consider the importance of considering the latest scientific evidence and ethical parameters when searching for and designing practical solutions to pressing environmental issues through emerging technologies. The chapter highlights the importance of combining systemic thinking with resilience thinking when approaching how technology can benefit the environment and communities, and transform economies according to a new paradigm of a greener and cleaner tech future based on circular, bio, and blue economic systems. It also aims at enhancing understanding on the relationship between business ethics and the environment, and the relevance of ethics and emerging technologies, including responsible research and innovation. It aims at reflecting on how our understanding of tackling environmental problems has influenced practical solutions for contemporary environmental problems. In particular, the chapter considers how the change in the paradigm will shape a new sustainable and greener future.

Background: Environment and sustainability

Over time, our understanding of tackling environmental problems has shifted as well as the drivers concurring with the success and failure of our endeavours. In the 1990s, the dominant trend indicated that socio-economic development objectives and environmental benefits could be achieved simultaneously according to adequate environmental and resources management. However, the scale of the world's resources exploration has reached a point where it has become increasingly difficult to achieve the so-desired Triple Bottom Line balance (Elkington, 2004, p. 23–24; Arruda, 2019) or the principles of sustainable development. Even considering that in different bioregions the perception of sustainable development may vary, as a global community we have realized the drastic signs that nature is sending to humanity. The level of complexity and interdependence of the Earth's systems revealed by the latest scientific evidence show that even solutions for problems can lead to unintended negative consequences in other geo-bioregions or trigger new socio-environmental problems in different parts of the globe.

In a fast-changing scenario, a rise in demand for natural resources is associated with the growth in the world's population and world GDP per capita. Human activity has expanded in terms of increased food production with consequences of soil depletion, wasting water, extermination of pollinators and wildlife, compromised genetic biodiversity, increased land use, decreased biochemical (nitrogen and phosphorus) flows, deforestation, *savanization* (Lovejoy and Nobre, 2018), and climate change. This increased demand for natural resources and correlated activities to attend the real and artificial needs of an increasing human population has proved not to respect the global system boundaries or the basic principles of environmental resilience (Arruda, 2018a, p. 112). Our current human systems do not respect the Earth's life-support capacity; in other words, our current patterns of production and consumption do not respect the resilience of the vital ecological processes needed to keep the planet fit for life (IUCN/UNEP/WWF, 1991).

This alarming statement/conclusion reveals the extent of the unbalanced relationship between the economic, environmental, social, and cultural pillars (Arruda, 2019, p. 65) of sustainability. We note with grave concern the consequences of continued global environmental degradation, including the scale of species decline (WEF, 2020) and a real threat to life on Earth irrespective of the bio-region of the world. Our current ways of dealing with pressing environmental problems have proved to be self-limiting because the extraction of natural resources represents on one side an opportunity but on the other hand a serious risk.

In previous decades, environmental issues were treated in silos without considering the economic and social impacts. Currently, with the development of interdisciplinary studies, environmental sciences and social sciences are dimensions of the same phenomena; understanding has emerged about the intrinsic relationship between environment and society. Solutions for the environment are also solutions for people and their livelihoods and well-being. Without societal responsibility, there is no environmental responsibility or solutions. Societal responsibility requires sustainability literacy and other literacies to enhance understanding of risks as well as the application of clear sustainable principles of a universal character.

Emerging local and global risks

In the latest study about global risks conducted by the World Economic Forum (2020) in partnership with business, academic, and public-sector communities, it was found that among the top ten risks by likelihood and impact over the next ten years, "climate-related issues dominated all of the top-five long-term risks by likelihood among members of the Forum's multistakeholder community" (see Figure 7.1). Aligned to this outcome, the World Economic Forum Global Risks Perception Survey 2019–2020 conducted among the members

Long-term risk outlook

Top 10 risks by likelihood and impact over the next 10 years

Multistakeholders

Likelihood	Impact
Extreme weather	Climate action failure
Climate action failure	Weapons of mass destruction
Natural disaster	Biodiversity loss
Biodiversity loss	Extreme weather
Human-made environmental disasters	Water crises
Data fraud or theft	Information infrastructure breakdown
Cyberattacks	Natural disasters
Water crises	Cyberattacks
Global governance failure	Human-made environmental disasters
Asset bubble	Infectious diseases

Figure 7.1 Top 10 risks by likelihood and impact over the next 10 years

Figure 7.1 illustrates the most important risks and impacts over the next ten years, providing a long-term outlook of effects on multistakeholders by likelihood and impact. This is an important figure calling attention to impacts on economic, environmental, and societal components of sustainability over the next 10 years.

Source: WEF (2020, p. 12)

of the Global Shapers Community represented by the youngest members of the Forum reinforced the ranking of environmental issues as the top risks in both the short and long term (see Figure 7.2).

These alarm signals sounded by scientists and climate experts appear to have been echoed by the frequency and scale of natural disasters and extreme weather events experienced by communities all over the world: "climate-related issues dominated all of the top-five long term risks by likelihood among members of the Forum's multistakeholder community" (WEF, 2020, p. 12). This perception is corroborated by individual and collective experiences of

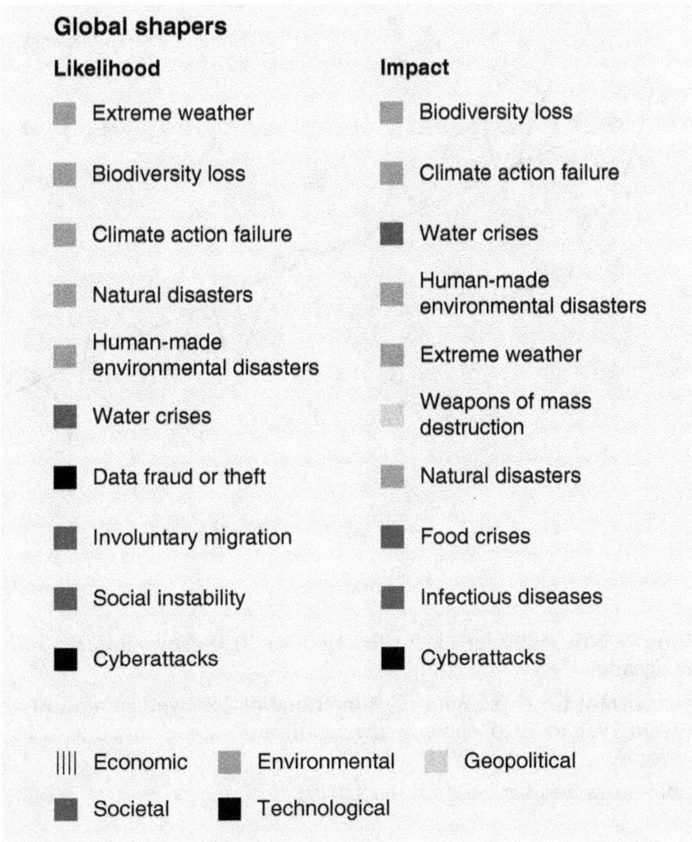

Figure 7.2 Top 10 risks by likelihood and impact over the next 10 years

Figure 7.2 illustrates the top ten risks by likelihood and impact over the next ten years based on economic, environmental, geopolitical, societal, and technological aspects. This illustrates the point on an integrated perspective to understanding risks and their impacts.

Source: WEF (2020, p. 12)

severe climatic variations with drastic impacts on well-being and livelihoods. Development factors represent an important element for management. Human health and food security are commonly affected by environmental components; consequently, there is significant interest in assessing risks caused by changes in climate, pollution, and environment.

In previous decades, the focus and perception of environmental challenges were associated with natural resources extraction (oil spills, mining residues) and environmental issues derived from industrial pollution, which motivated important case studies in the environmental management literature.

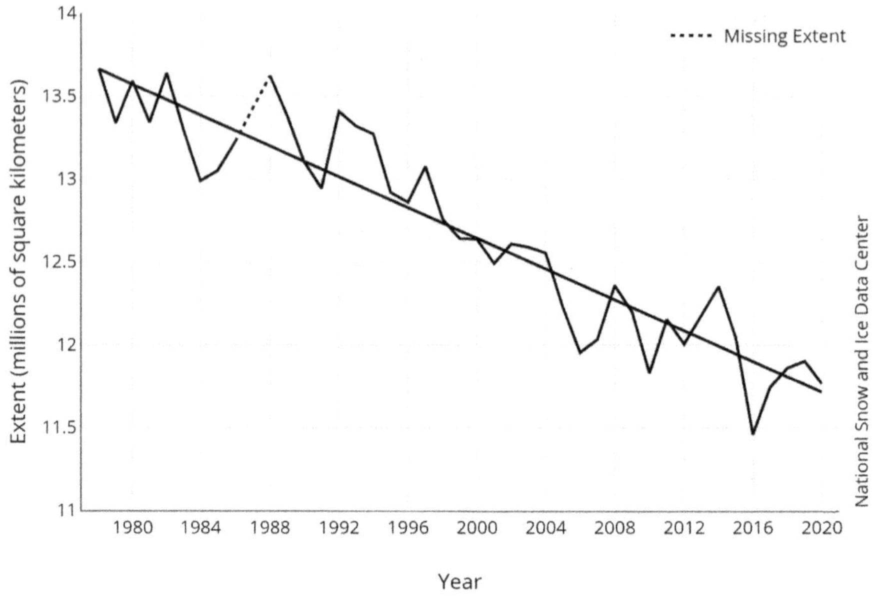

Figure 7.3 Monthly November ice extent for 1978 to 2020 shows a decline of 5.1% per decade

Figure 7.3 is important for explaining the timeframe of the average monthly Arctic sea ice extent from 1978 to 2020, showing the significant sea-ice retreat in millions of square kilometres.

Source: National Snow and Ice Data Center (2020)

Real-world problems were fundamentally reduced to localized environmental issues purely addressed by technology applied within environmental management strategies as if only technology would solve them all at a local level. Currently, the emergency of climate change has translated into multi-scale geographical events impacting communities around the world. This shows the complexity of the climate regime and the interdependent layers of complex dynamic interactions whose impacts are unpredictable and still unknown.

The Arctic plays a key role in maintaining the balance of bio-geophysical dynamics and the stability of the global climate. As the ice retreats, current challenges concerning the environment, maritime trade routes, tourism, oil and gas activity, and other extractive activities will further intensify the effects of climate change on ecosystems and communities (Arruda, 2015, 2018b). The climate impacts on the Arctic are than just temperature and precipitation, but relate to a complex system of snow, ice, and circulation patterns in the atmosphere and oceans. Sea ice is one of the most important climatic variables and a key agent and indicator of climate change. It also affects surface reflectivity, cloudiness, humidity, exchange of heat and moisture, the ocean surface, and oceans' currents (AMAP, 2004). Sea ice has a profound effect on Arctic temperatures. Ice has a bright, reflective surface that helps beam sunlight away from the earth. Thick sea ice also helps insulate the ocean, trapping heat below the surface in the winter and preventing it from escaping into the cold Arctic air. As the sea ice retreats, the ocean absorbs more heat in the summer, and in the winter that heat escapes through the thinner ice and warms up the atmosphere. The rising temperatures speed up the transition from snow to rain.

The latest polar data based on the NSIDC Sea Ice Index (Fetterer et al., 2017) about sea-ice retreat shows a concerning estimate that the 2020 Arctic sea ice retreat season was marked by extreme conditions with persistently low averages in September and October 2020. The total sea-ice extent for November 2020 was the second lowest in the satellite record and in December "sea ice extent remains far below average, dominated by the lack of ice on both the Pacific and Atlantic sides of the Arctic Ocean. As was the case for October, air temperatures averaged for November were well above average over much of the Arctic Ocean, notably over open water areas" (Meier et al., 2020).

The Arctic region is experiencing the results of climate change with diminishing ice, permafrost melting, erosion, and several other negative consequences. Environmental data and the analysis of environmental indicators should be considered together with data on demographic and societal changes in the region. It is important to have holistic and specific strategies and plans for climate change mitigation that take into consideration all aspects of sustainable development.

It is important not to forget that sea ice is only one variable in this complex equation. There are a myriad of other effects happening simultaneously on land, in the atmosphere, and in the intricate network of ecosystems. Climate change is transforming the Arctic environment at a speed that will affect local and global societies before they completely understand the interconnected physical, social, and environmental alterations underway. As per the fast pace of change experienced, risk assessment and risk management require an innovative, integrative, and coordinated local–global approach of risk analysis to identify mechanisms of resilience for the region.

Ethics and emerging technologies: Responsible Research and Innovation

Groups have become skilled at addressing conventional environmental risks – risks that can be easily identified and locally managed through standard risk management approaches. However, we are living in unprecedented times when environmental risks are not completely predictable and assessable but are defined by complexity and tipping points. There are transdisciplinary risks with a range of effects on economic, social, cultural, and geographical aspects. These are times when transformational change is needed at a systemic level. While it is important to employ various technologies to reduce industrial emissions and mitigate climate change and minimize environmental impacts, it is also essential to ensure comprehensive and systemic strategies and solutions to address the present and the future of the Arctic region as a whole. Integration of social, environmental, and ethical concerns into business operations and core strategy in dialogue with stakeholders needs to become mainstream practice.

Technological leadership is a landmark of the Arctic. There are several high-technology businesses originated in the Arctic that address local but global challenges. The technological leadership of these companies is a result of a long-term focus on research, development, and spirit of innovation to design practical solutions to pressing problems. There are important clusters of innovation in the Arctic. The term cluster is used "to refer to a number of organizations sharing similar aims, characteristics, goals, and activities with a shared sense of geographical belonging" (Middleton et al., 2019, p. 23). This sense of belonging, collaboration, and specialization allows these companies "to capitalize on nature-related resources in a creative way" (Middleton et al., 2019, p. 23) and develop efficient solutions for real problems based on their efficient local networks that are sometimes small but unique in their highly technical infrastructure and knowledge base. The main features of the Arctic clusters of innovation are:

- Clusters that are rather small but many in number.
- Cooperation concepts: circular economy, commercialization of knowledge, technological leadership, integration of supply/value chains.
- Empowered by close-knit local communities.

(Middleton et al., 2019, p. 24)

Important clusters operate in fields like bioeconomy, biotechnology, renewable energy, bio-marine resources, digitalization, health care, bio-medicine, hospitality and entertainment, seafood, food and drink, and aquaculture. The Nordic clusters operate according to organic growth and local ownership, and are based on cooperation concepts of the circular economy, technology development, technological leadership, integration, shared values of quality, uniqueness, and excellence. The clusters are based on applied science/knowledge but also applied ethics.

Partnerships for research and development are also the base of successful technologies. Ansvarlig Forskning og Innovasjon i Norge (AFINO) is a network and learning centre for responsible innovation and corporate social responsibility (CSR) in Norway, consisting of associated research partners that jointly explore responsible innovation to address environmental and societal challenges. This initiative provides a better integration between research, industry, community, and policy-making (NTNU, 2020a, b). It captured the necessity of creating research and resource-building virtual clusters of Responsible Research and Innovation (RRI) and CSR to critically discuss and address science, ethics, social science, innovation, and the respective interactions between science, social sciences, business ethics, cross-sectoral learning, and policy-making in an interdisciplinary and applied way according to concrete implementation orientation. This is an ongoing project using a virtual platform to promote interaction and a more integrative approach among authorities, companies, civil society, and policymakers to investigate how different industries and institutions can contribute to a more sustainable society by implementing social responsibility in new ways (OsloMet, 2020). This interdisciplinary platform for network and discussion about RRI and CSR is funded and coordinated by the Norwegian Research Council Responsibility Program and inspired by previous activities and achievements regarding Biotechnology for Innovation (OECD, 2020), the Research Programme on Nanotechnology and Advanced Materials (NANO2021), the Initiative for ICT and Digital Innovation (IKTPLUSS), and the Programme on Responsible Innovation and Corporate Social Responsibility (SAMANSVAR).

NANO2021 is a large-scale research programme on nanotechnology, microtechnology, and advanced materials aiming at advancing knowledge and technology application in these fields towards renewable energy, reduced negative impacts on the environment and climate, improved health and medical technology, and sustainable utilization of natural resources (RCN, 2006, 2018). Research on impacts of nanomaterials on health and ecosystems is also a thematic priority area (NANO2021, 2020). The objective of this strategic programme is to promote the responsible development of nanotechnology to contribute to industrial and commercial development that will be of relevance to environmental solutions and benefit society. The practical applications through interdisciplinary cooperation and innovation in the industrial sector address environmental challenges related to clean-tech energy (ENERGIX), petroleum research (PETROMAKS 2), and food and bio-based industries (BIONÆR).

These are practical initiatives in need of continuous alignment, tuning, and development in the academy and industry, based on the dimensions shown in Table 7.1.

Circular economy

Sustainable development is not only about producing through clean tech but also about changing patterns of production and consumption. It is about producing less and using fewer resources by exerting responsibility in eco-innovation, resource efficiency, and waste management. It consists in changing the paradigm from the "traditional linear economic model based on a 'take-make-consume-throw away' pattern", to an economy "based on sharing, leasing, reusing, repairing, refurbishing and recycling" where products and the materials they contain are highly valued and reused by reducing waste to a minimum (Bourguignon, 2016, p. 1).

Table 7.1 Dimensions of responsible innovation. Based on 'A Framework for Responsible Innovation' – under BIOTEK2021, IKTPLUSS, NANO2021, and SAMANSVAR

Anticipatory	Technologies may potentially have terrain-changing effects that are realized in complex, dynamic interplay with other societal forces. There is lack of knowledge and understanding concerning how to realize this potential in relation to desired societal development.
Reflexive	This involves employing expertise and capacity to better identify and discuss prerequisites for research and innovation activities, in the form of fundamental, often implicit, assumptions and frameworks of understanding, irreducible uncertainty, and areas of ignorance. A greater degree of reflexivity is vital in order to provide directionality in research and innovation processes.
Inclusive	To "enable society to speak back to science", attention is now increasingly being directed towards the research environments themselves. What is being targeted are the skills needed to open up research and innovation processes, recognize the limits of one's own knowledge and competence, and the ability to ask for help in dealing with the potentially landscape-changing effects of these processes.
Responsive	Activities involving the first three dimensions are intended to provide continual input and substance to new governance practices. This entails the development of horizontal or distributed governance schemes that encourage collaboration with partners that may be affected by a research and innovation process. There is a need to open up different perspectives relating to dilemmas and irreducible uncertainty. This must take place via broad-based involvement, not only on the part of researchers from different disciplines, but also bringing on board policy actors, including research councils, trade and industry, interest organizations, and society at large. The RRI method is a learning process with no fixed answers ("beyond rules and regulations").

Sources: Norwegian Board of Technology and Research Council of Norway (2015); Strand et al. (2015)

The circular economy is an economic model based on this principle of retaining the highest utility and value of products, components, and materials with an extended producer responsibility (EPR) for taking back used goods and sorting their recycling (Bourguignon, 2016, p. 2). The main argument in defence of a more circular economy is to reduce pressures on the environment based on principles of biological cycles, because there are biological materials which can be decomposed by living organisms and 'technical' materials which cannot be decomposed by living organisms. The responsibility principle is highly amplified in this concept because the "the aim is to limit the leakage of resources as much as possible" (Bourguignon, 2014). According to data from the European Commission, it is estimated that over 600 million tonnes of CO_2-equivalent emissions by 2035 (on average 30 million tonnes per year) would be avoided by the adoption of the circular economy package. A reduction of EU CO_2-equivalent emissions by 450 million tonnes per year until 2030 (EU Commission, 2015) could be achieved with the leading position to be climate neutral by 2050.

The circular economy represents a way forward as the impact of climate change would potentially be minimized by CO_2 emissions halving by 2030 compared to today's levels, representing 48% reduction of carbon dioxide emissions by 2030 or 83% by 2050. Apart from the clear minimization of climatic impacts and environmental protection, the circular economy is potentially important for enhancing the security of supply of raw materials; increased competitiveness; innovation; economic growth; and boosting the green economy and a new generation of green jobs. However, the most remarkable effect of this coordinated action is mapping, monitoring, and practically integrating several aspects of production, consumption, waste management, raw materials, climate neutrality, economics, markets, value digitalization, research & development, and value chains. This will enable circularity to work for people, regions, and cities, in order to monitor progress and lead efforts at the local and global level.

Smart specialization

The Arctic region with its unique characteristics of sparsely populated communities, a vast land area, and wealth in natural resources presents a range of opportunities but also additional challenges related to climate change adaptation not seen elsewhere in the world. The Arctic region is developing at a fast pace, taking the lead in several industries, such as forestry and bioeconomy, mining and metallurgy, and fisheries in areas of the European Arctic like northern Finland, northern Sweden, and northern Norway. Smart specialization has supported regions in addressing Arctic challenges through a collaborative approach, joint investment, and a spirit of endurance using pioneering technologies. Although it is a region that is heavily dependent on traditional activities like fishing, hunting, forestry, and mining, it is well known in other economic sectors such as bioeconomy, agriculture, creative industries, tourism, ICT and R&D, cold-climate technologies, maritime activities, and blue growth (Teräs et al., 2018, p. 7).

Smart specialization is a place-based approach aiming at prioritizing public research and innovation investments in priority domains for the economic transformation of regions. It is based on the strengths of specific economic contexts (new combinations between region-specific capacities and region-specific opportunities) (Foray et al., 2018), building on regional competitive advantages (Barca, 2009) to discover new niches from cross-sectoral synergies, open new growth paths, and diversify the economy into new domains of specialization (Crespo et al., 2017). At the core of the smart specialization strategy is discussion about sustainable growth and sustainable development

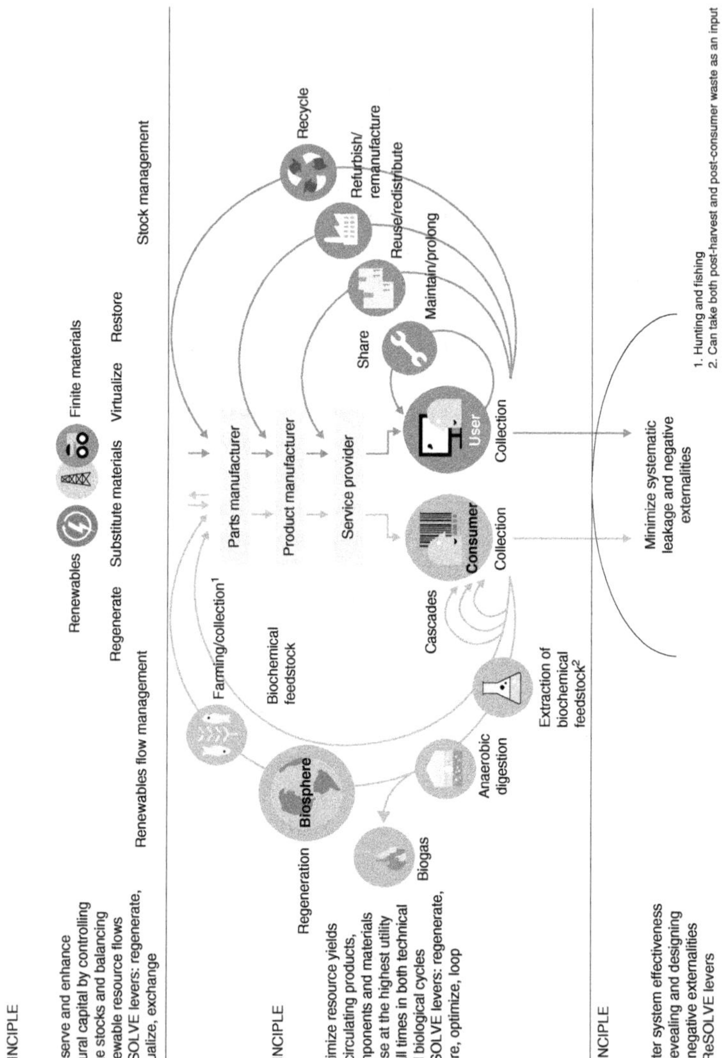

Figure 7.4 Outline of a circular economy

Figure 7.4 is important for outlining the principles of the circular economy based on natural capital, optimization of resources, and systems effectiveness. Important aspects of regeneration, virtualization, and exchange towards the minimization of negative externalities are illustrated in this figure.

Source: Ellen MacArthur Foundation (2015, p. 24)

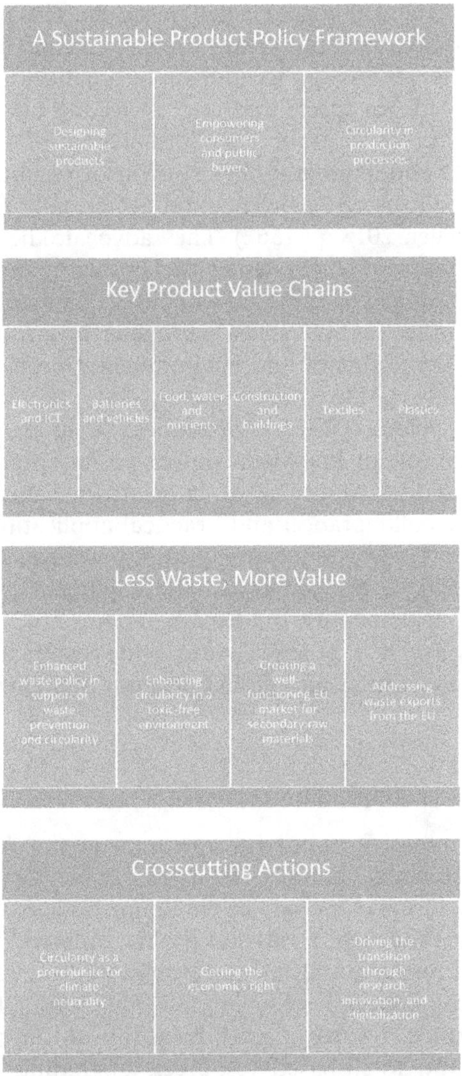

Figure 7.5 A sustainable product policy framework

Figure 7.5 presents several aspects of a sustainable product policy framework based on key product value chains, circularity in production processes, and enhanced waste policy. It shows the several interconnected aspects of an operative sustainable product policy.

Source: Adapted from Ellen MacArthur Foundation (2015)

(European Commission, 2010), increasing the role of high technology but not being limited to it to address the overall changes needed in society. This seems to contribute to an intensified discussion about pressing environmental issues like climate change. However, it is not enough to address, in practical terms, the impacts of the interactions between specialized-style firms and old-style traditional firms. There are authors who argue that "the process of transformative change based on R&D and national policy systems of technology and innovation are unfit for addressing environmental issues and social challenges" (Schot and Steinmueller, 2018, p. 1562). They advocate the need to focus on innovation and development at a systemic level but with practical application towards sustainability. It can be achieved with high alignment and adherence of smart specialization to the SDGs as a parameter to safeguard the balance between economic development and environmental protection (Arruda, 2019).

The practical application of technological opportunities, i.e. in environmental contexts, could unveil the role of knowledge-intensive entrepreneurs (Malerba et al., 2016; Malerba and McKelvey, 2018) in smart specialization in order to create new modes of collaboration and practical application by valuing the opportunities created by diverse technological knowledge, skills, and levels (low, mid or high tech) to provide systematic alignment to the SDGs and therefore accomplish sustainability. Innovation systems need to be sustainable in a way that are applicable in practice and provide sustainability to old structures in order to support local and global business and environmental solutions.

Bioeconomy: Green transition, principles, and criteria

Given the large-scale resource-intensive industries operating in the Arctic (multiple commodities such as base metal mining, oil and gas), when focusing on responsibility and practical solutions for complex environmental contexts, it is important to adopt systemic models of multi-stakeholder collaboration. In these models, knowledge and resource sharing can benefit all by creating circular and self-sustainable local solutions following re-evaluation of local value chains. These are transformed towards a green transition through the replacement of unsustainable and fossil-based resources, upgrading side-streams and waste (Nordic Council of Ministers, 2017, p. 73) to achieve zero-waste, zero-emissions, and the core values of sustainable development. The developed bioeconomy sector in the European Arctic provides a powerful policy instrument for responsible regional economic development while addressing negative environmental impacts.

This is possible for three specific reasons: the principles intrinsic to bioeconomy, efficient technologies, and coordinated practices that integrate a number of solutions by balancing the environmental, social, and economic pillars of sustainability. According to The Nordic Council of Ministers (2017):

> The bioeconomy consists of the management of renewable biological resources and their conversion into food, livestock feed, bio-based products and bioenergy via innovative and efficient technologies. It means utilising biomass to its maximum sustainable potential in terms of both volume and value – in other words, using every part of the biomass and using it intelligently. The bioeconomy covers all kinds of products: energy, biofuel, heat, construction, bioplastics, smart packaging materials, food, livestock feed, ingredients, textiles, health and pharmaceuticals, just to name few.

Bioeconomy is a holistic circular and coordinated approach that recognizes the support capacity of the planet, the challenge of climate change, as well as the strengths of environmental and social resilience (Arruda and Krutkowski, 2017, p. 279) based on the following principles:

- Replace – in search of responsible resources
- Upgrade – unlocking the full potential of side-streams and waste
- Circulate – creating self-sustainable, local, and circular bio-solutions
- Collaborate – because we are smarter together

In the Arctic, energy, water, and food are inseparable components of activities and practices to replace fossil-based resources with bio-based resources. There is significant potential for a land-based bioeconomy (forestry) in the Arctic areas of Sweden and Finland. In northern Norway, Sweden, and Finland, reindeer husbandry is an economic activity that is strongly linked to the Sámi indigenous population and an inspiring example of traditional activity (Arruda and Krutkowski, 2016) that has lasted for thousands of years of Sami sustainable practices.

Blue economy

In this section, we approach the blue economy concept to again emphasize the relevance of an integrated understanding about the functions of environmental assets, ecosystem services, and socio-environmental systems from the viewpoint of present and future use, management, and resilience. The main factor under discussion is the sustainability core criteria as the conceptual model should provide social and economic benefits for current and future generations; restore, protect, and maintain marine ecosystems; and be based on clean technologies, renewable energy, and circular material flows (WWF, 2016). In this sense, a particular useful conceptualization of the term 'blue economy' is the use of water resources (aquatic resources – marine, rivers, lakes) to increase socio-economic and human well-being, while reducing environmental risks and ecological scarcities (Smith-Godfrey, 2016, p. 2).

The blue economy concept emphasizes the importance of valuing the marine (aquatic) economy and how industrial activities, transportation, livelihoods,

Table 7.2 Bioeconomy principles

Replace	In search of responsible resources	Economy based on plant- and animal-based resources (bio-resources).	When produced in a sustainable manner, these materials have many beneficial effects, including carbon capture through excess photosynthesis, curtailing erosion, protecting the soil, providing habitats for wildlife, and many other ecosystem services.
Upgrade	Unlocking the full potential of side-streams and waste	Optimization throughout the whole value chain – utilizing the full potential of raw materials that are caught and harvested – efficiency.	Increasing the value of currently unused and underutilized living natural resources, as well as unlocking the full potential of residues and waste.
Circulate	Creating self-sustainable, local and circular bio-solutions	The circularity of the bioeconomy must take into account the sustainability of the ecosystems and their ability to replace the annual harvest of biological material. The regeneration processes must not entail significant negative externalities, e.g. pollution, loss of biodiversity, or adverse climate effects.	The circular bioeconomy's ecosystem-based approach requires extensive scientific knowledge about the functions of the different ecosystems, as well as the implementation of surveillance systems to monitor their status and development.
Collaborate	Because we are smarter together	High level of social capital and trust – collaborative structures.	Collaboration is crucial for the successful development of the Nordic bioeconomy's other three pillars: replace, upgrade, and circulate. Companies in different sectors must align their processes in order to optimize resource utilization; R&D institutions must work closely with the private sector to provide innovative solutions, and regional authorities must collaborate with industry to ensure the right infrastructure is in place and the appropriate incentives are offered.

Source: Nordic Council of Ministers (2017)

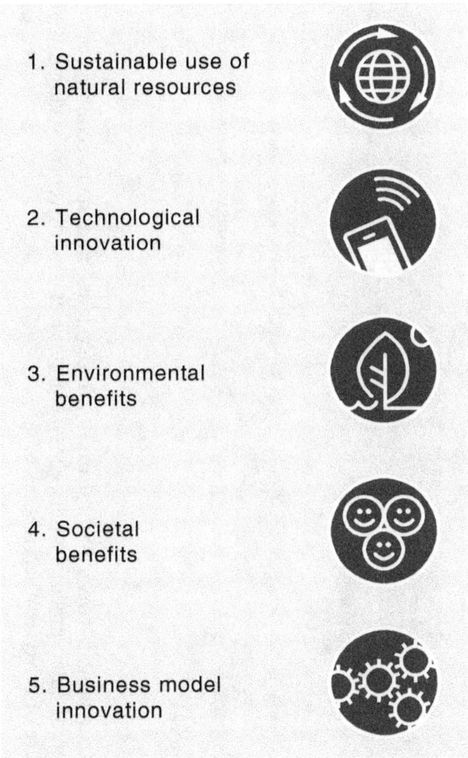

Figure 7.6 Sustainable and innovative bioeconomy criteria

Figure 7.6 presents the criteria for a sustainable and innovative bioeconomy based on natural resources, technology, environmental and societal benefits, and new business models. It is a relevant figure suggesting the connections among different aspects of bioeconomy in the North.

Source: Nordic Council of Ministers (2017)

food security, recreational activities, and ecosystems are interlinked, particularly when climate change is altering the sea ice volume and opening new maritime routes in the Arctic. The question that remains is how these activities will accomplish the core principle of blue economy concept (European Commission, 2020a): reduce environmental risks and ecological scarcities. Water, energy, and food are closely interrelated resources and need to be considered simultaneously, along with maintaining the biophysical resource base – land, soil, hydrological, and biological resources – to ensure the sustainable delivery of ecosystem services.

The Arctic changing environment and the material impact on Arctic ecosystem services not only threatens its blue resources but also compromises communities' subsistence and economies, health, and well-being. Sea-ice

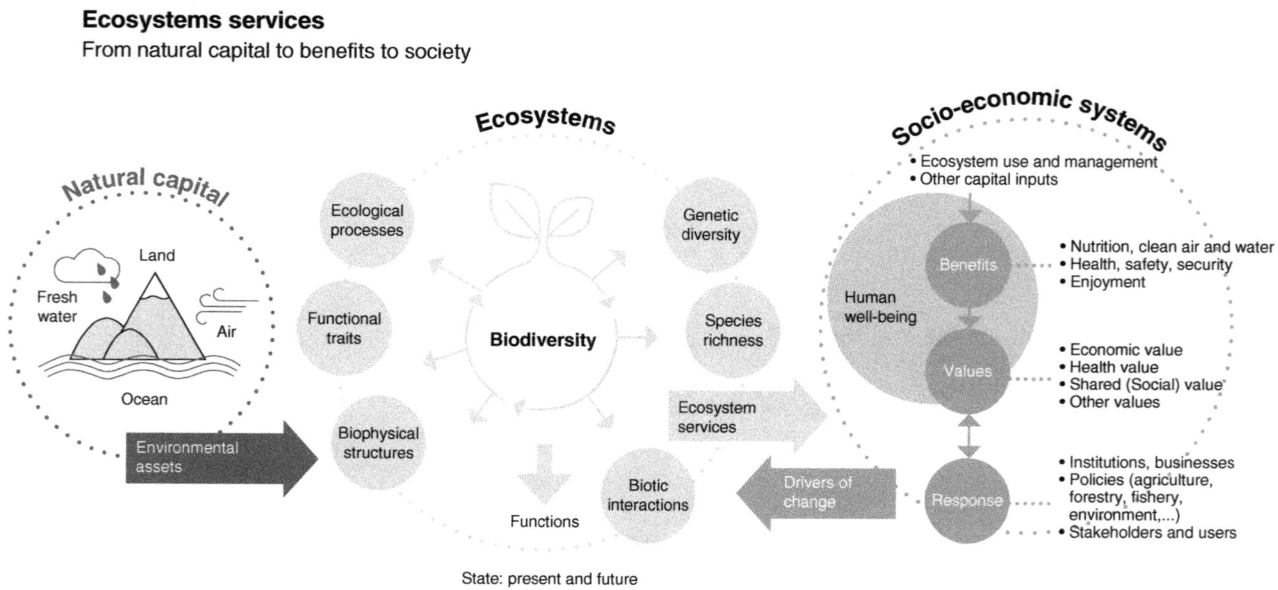

Figure 7.7 Ecosystem services

Figure 7.7 is especially important for outlining the interactions between natural capital, ecosystems, and socio-economic systems. It clearly explains the relevance of the interaction of environmental assets, ecosystem services, and drivers of change. It provides an important outlook of the present and future of ecosystem services.

Source: European Commission (2020, p. 50)

retreat, rising ocean levels and temperature, ocean acidification, and thawing permafrost affect and change food web dynamics. Additionally, other risks are intensified such as increased exposure to severe weather events, coastal erosion, infrastructure damage, risky subsistence activities (hunting, fishing), and increasing ice-related injuries and deaths (Kathijotes, 2013; Jensen, 2018). These challenges are the signals of emergent and complex environmental contexts in need of prompt coordinated action, putting concepts and models into practice to sustain a resilient, diversified, and durable economy. The maritime industry, universities, and fishermen's associations all over the world (i.e. Iceland, Norway, Canada, Ireland) are practising the cluster concept discussed in previous sections of this chapter by engaging public, private, and industry stakeholders to increase collaboration among business, government, and non-profit organizations, enhancing opportunities for blue economy innovation and entrepreneurship and strengthening relations with other Arctic nations.

Further resources

Several other environmental contexts could be approached using the guidelines and alignments of this chapter, e.g. energy, trade routes, and pollution.

In *Renewable Energy for the Arctic: New Perspectives* (Arruda, 2018b), there is a fruitful discussion about the challenges, opportunities, and trends of a low-carbon energy system and the unique position of the Arctic region in leading Arctic stakeholders to adopt initiatives to decrease fossil fuel use and consequently GHG emissions. The book provides new socio-environmental and intercultural perspectives on the Arctic energy landscape, discussing innovative energy solutions in Greenland, Iceland, Svalbard, Canada, and Alaska, such as the glacial meltwater power project; geothermal energy development; the integrated MARKAL-EFOM System; smart grids to integrate energy infrastructure to benefit communities, corporations, and the renewable energy transition; and management and the implications of Nordic electricity interconnection.

The Nordic Council of Ministers (2017) has published a report including "25 Nordic cases on the 'new' Nordic bioeconomy". This is an interesting source of complementary materials to expand understanding on the management of renewable biological resources.

The progressive and irreversible transition to a sustainable economic system (Köhler et al., 2019) is the core plan of the European Union in its Circular Economy Action Plan and is accelerating the transformational change aimed at by the European Green Deal, while building on previous circular economy actions. In *Circular Economy Action Plan: For a cleaner and more competitive Europe* (European Commission, 2020b), innovative models based on a collaborative economy powered by innovative technologies intend to accelerate circularity in a systemic, deep, and transformative way through alignment and cooperation of all stakeholders at all levels: local, national, regional, and international.

Case study

The concept of bio-refineries

Among the criteria of a sustainable bioeconomy, the replacement of petroleum-based products and additives can mean a real step forward in promoting a sustainable economy. The case for bio-based additives from the Exilva plant in Norway is a remarkable example in this regard. Exilva is a wood-based performance-enhancer that replaces petroleum-based additives in several different applications like adhesives, coatings, agricultural chemicals, and cosmetics. Exilva consists of a network of suspended microfibers known as micro-fibrillated cellulose extracted from wood and converted into a network of microfibrils. It is a multifunctional additive, which reinforces and stabilizes various substances. Its bio-based material (cellulose from wood) is made from 100% natural raw materials and represents the most abundant organic polymer in nature. It is not only the plant-based raw material that makes it a sustainable option of a higher-performance additive as a substitute for petrochemical-based synthetic alternatives, but the process of production based on renewable energy has the potential of opening new market niches as well as green job opportunities.

Questions for discussion

1. How can technology benefit the environment and communities?
2. What are the principles of the circular economy?
3. What is the concept of smart specialization?
4. What is importance of the blue economy?
5. What are the principles of the bioeconomy?
6. What are the criteria for a sustainable and innovative bioeconomy?

Chapter summary

In the Earth's system, nothing lives in isolation. This chapter highlighted the importance of systemic thinking and resilience thinking when approaching how technology can benefit the environment and communities, and transform economies according to a new paradigm of a greener and cleaner tech future. It also emphasized the importance of considering the latest scientific evidence and durable parameters when searching for and designing practical solutions to pressing environmental issues through emerging technologies.

Responsibility is not a new idea; however, it needs to evolve and revitalize itself over time due to the changing features of business, society, and the environment. CSR in its multitude of concepts and interpretations can be seen and practised when risks are precisely evaluated, when clusters of Responsible Research and Innovation exercise collaboration and specialization allowing companies to capitalize on nature-related resources in a creative and sustainable way. Several new economic models have been designed in an attempt by businesses to balance and integrate their economic, social, and environmental responsibilities in a way that minimizes societal harm and optimizes societal benefit while providing wealth to business owners and shareholders (Webb, 2014). The eternal dilemma is how to promote this fine balance in complex and unpredictable contexts. The recent range and scale of pressures on businesses, related to the effects on environmental conditions and communities, however, has made organizations rethink responsibility from different perspectives and in the long term. The pressure for CSR to address complex environmental scenarios with critical ramifications is becoming stronger because businesses have realized that the challenges are gigantic, complex, and interconnected. Time is of the essence to prevent failures in addressing the even more exigent environmental targets, because in the end it is not a matter of standards but survival.

References

Alaska Coastal Community Protection Project. (n.d.). Alaska State Department of Commerce, Community, and Economic Development. Accessed on 15 November 2020 from www.commerce.alaska.gov/web/dcra/PlanningLandManagement/AlaskaCommunityCoastalProtectionProject.aspx

Arctic Monitoring and Assessment Programme (AMAP). (2004). *Impacts of a Warming Arctic*. Cambridge, UK: Arctic Climate Impact Assessment (ACIA).

Arruda, G.M. (2015). Arctic governance regime: The last frontier for hydrocarbons exploitation. *International Journal of Law and Management*, 57(5), 498–521.

Arruda, G.M. (2018a). Arctic resource development: A sustainable prosperity project of co-management. In G.M. Arruda (Ed.), *Renewable Energy for the Arctic: New Perspectives*. Abingdon, UK: Routledge.

Arruda, G.M. (2018b). *Renewable Energy for the Arctic: New Perspectives*. Abingdon, UK: Routledge.

Arruda, G.M. (2019). *Sustainable Energy Education in the Arctic: The Role of Higher Education* (1st ed.). Abingdon, UK: Routledge.

Arruda, G.M. and Krutkowski, S. (2016). Arctic governance, indigenous knowledge, science and technology in times of climate change: Self-realization, recognition, representativeness. *Journal of Enterprising Communities: People and Places in the Global Economy*, 11(4), 514–528, p. 519.

Arruda, G.M. and Krutkowski, S. (2017). Social impacts of climate change and resource development in the Arctic: Implications for Arctic governance. *Journal of Enterprising Communities: People and Places in the Global Economy*, 11(2), 277–288.

Barca, F. (2009). *An Agenda for a Reformed Cohesion Policy*. Brussels: DG REGIO.

Bourguignon, D. (2014). *Turning Waste into a Resource: Moving Towards a 'Circular Economy'*. Briefing December 2014. Brussels: European Parliament Research Service (EPRS), European Parliament.

Bourguignon, D. (2016). *Closing the Loop: New Circular Economy Package*. Briefing January 2016. Brussels: European Parliamentary Research Service, European Parliament.

Crespo, J., Balland, P., Boschma, R., and Rigby, D. (2017). *Regional Diversification Opportunities and Smart Specialization Strategies*. Brussels: European Commission, Directorate-General for Research and Innovation, Publications Office of the European Union. doi: 10.2777/133737

Ellen MacArthur Foundation. (2015). *Growth Within: A Circular Economy Vision for a Competitive Europe*. Ellen MacArthur Foundation, SUN and McKinsey Center for Business and Environment.

Elkington, J. (2004). Enter the triple bottom line. In A. Henriques and J. Richardson (Eds.), *The Triple Bottom Line: Does It All Add Up? Assessing the Sustainability of Business and CSR* (pp. 1–16). London: Earthscan Publications Ltd. https://doi.org/10.1108/13598540910941948.

European Commission. (2010). *Regional Policy Contributing to Smart growth in Europe 2020*. COM(2010)553. Brussels: European Commission.

European Commission. (2015). Factsheet: An ambitious EU circular economy package. Accessed on 25 April 2020 from https://ec.europa.eu/commission/publications/ambitious-eu-circular-economy-package_en

European Commission. (2020a). *The EU Blue Economy Report 2020*. Luxembourg: Publications Office of the European Union.

European Commission. (2020b). Communication from the Commission to the European Parliament, the Council, the European Economic and Social Committee and the Committee of the Regions. *A New Circular Economy Action Plan: For a Cleaners and More Competitive Europe*. COM/2020/98 final.

European Community. (2010). *Europe 2020: A European Strategy for Smart, Sustainable and Inclusive Growth*. Luxembourg: EU.

Fetterer, F., Knowles, K., Meier, W., Savoie, M., and Windnagel, A.K. (2017). Sea Ice Index, Version 3. Boulder, CO: NSIDC National Snow and Ice Data Center. https://doi.org/10.7265/ N5K072F8

Foray, D., Morgan, K., and Radosevic, S. (2018). *The Role of Smart Specialisation in the EU Research and Policy Landscape*. European Commission. Accessed on 20 June 2020 from https://ec.europa.eu/regional_policy/en/information/publications/brochures/2018/therole-of-smart-specialisation-in-the-eu-research-innovation-policy-landscape

Gifford, E. and McKelvey, M. (2019). Knowledge-intensive entrepreneurship and S3: Conceptualizing strategies for sustainability. *Sustainability*, 11, 4824.

Helm, D. (2007). *The New Energy Paradigm*. Oxford, UK: Oxford University Press.

International Union for Conservation of Nature/United Nations Environment Programme/World Wildlife Fund for Nature (IUCN/UNEP/WWF) (1991). *Caring for the Earth: A Strategy for Sustainable Living*. Gland, Switzerland: Author.

Jensen, A. (2018). Blue economy is future of Alaska economy. Accessed on 5 April 2019 from Alaska Journal website: www.alaskajournal.com/2018-11-28/blue-economy-future-alaska-economy

Kathijotes, N. (2013). Keynote: Blue Economy – environmental and behavioural aspects towards sustainable coastal development. *Procedia – Social and Behavioral Sciences*, 101, 7–13.

Köhler, J., Geels, F., Kern, F., Markard, J., Onsongo, E., Wieczorek, A., Alkemade, F., Avelino, F., Bergek, A., Boons, F. et al. (2019). An agenda for sustainability transitions research: state of the art and future directions. *Environmental Innovation and Societal Transitions.*, 31, 1–32.

Lovejoy, T.E. and Nobre, C. (2018) Amazon tipping point. *Science Advances*, 4, eaat2340.

Malerba, F., Caloghirou, Y., McKelvey, M., and Radosevic, S. (2016) (Eds.) *Dynamics of Knowledge-Intensive Entrepreneurship: Business Strategy and Public Policy*. Abingdon, UK: Routledge.

Malerba, F. and McKelvey, M. (2018). Knowledge-intensive innovative entrepreneurship integrating Schumpeter, evolutionary economics, and innovation systems. *Small Business Economics*, in press.

Meier, W., Bhatt, U.S., Walsh, J., Thoman, R., Bieniek, P., Bitz, C.M., Blanchard-Wrigglesworth, E., Eicken, H., Hamilton, L.C., Hardman, M. ... et al. Editors: B. Turner-Bogren, H.V. Wiggins, and S. Staudt. (2020). *2020 Sea Ice Outlook Interim Post-Season Report*. Accessed on 20 December 2020 from www.arcus.org/sipn/sea-ice-outlook/2020/interim

Middleton, A., Hersinger, A., Bryksenkov, A., Mineev, A., Dybtsyna, E., Bullvåg, E., Simonen, J., Pesämaa, O., Dahlin, P., and Ovesen, S. (2019). Business Index North – A periodic report with insight to business activity and opportunities in the Arctic. *Business Index North*, issue 3, vol. June 2019.

NANO2021. (2020). Accessed on 5 January 2021 from www.forskningsradet.no/en/about-the-research-council/programmes/nano2021/)

Norwegian University of Science and Technology (NTNU). (2020a). AFINO. Accessed on 5 January 2021 from www.ntnu.edu/afino/the-afino-research-school

Norwegian University of Science and Technology (NTNU). (2020b). AFINO. Accessed on 5 January 2021 from www.ntnu.edu/afino/about-afino

National Snow and Ice Data Center. (2020). Sea Ice Index. Accessed on 10 January 2021 from https://nsidc.org/arcticseaicenews/2020/12/persistently-peculiar/

Nordic Council of Ministers. (2017). Nordic bioeconomy – 25 cases for sustainable change. Accessed on 13 July 2020 from www.norden.org/en/news/25-nordic-bioeconomy-cases-sustainable-change

Norwegian Board of Technology and Research Council of Norway. (2015). 21-prosessenes samfunnsansvar (Societal responsibility in the "21" R&D strategy processes for the 21st century). Accessed on 20 December 2020 from www.forskningsradet.no/contentassets/558d5b1a9f53421f81371ecf96cf1692/framework-responsible-innovation.pdf

Organisation for Economic Co-operation and Development (OECD). (2020). BIOTEK 2021. Biotechnology for Innovation. Accessed on 5 January 2021 from https://stip.oecd.org/stip/policy-initiatives/2019%2Fdata%2FpolicyInitiatives%2F5653

OsloMet. (2020). AFINO. Accessed on 5 January 2021 from www.oslomet.no/forskning/forskningsprosjekter/afino

Owen, R., Macnaghten, P., and Stilgoe, J. et al. (eds.) (2013). *Responsible Innovation: Managing the Responsible Emergence of Science and Innovation in Society*. "Responsible Research and Innovation" – report to ERIAB (March 2014), UK.

Research Council of Norway (RCN). (2018). Large-scale Programme Nanotechnology, Microtechnology and Advanced Materials – NANO2021. The Research Council of Norway. Accessed on 25 January 2021 from www.rcn.no/nano2021

Research Council of Norway (RCN). (2006). A good council? Evaluation of the Research Council of Norway. Technopolis (2012); Egenvurdering av satsingen på foresight og dialogbaserte arbeidsformer i Norges forskningsråd 2003–2005 (Internal assessment of the Research Council's foresight initiative).

Schot, J. and Steinmueller, W.E. (2018) Three frames for innovation policy: R&D, systems of innovation and transformative change. *Res. Policy*, 47, 1554–1567.

Smith-Godfrey, S. (2016) Defining the blue economy. *Maritime Affairs: Journal of the National Maritime Foundation of India*, 12(1), 58–64. doi:10.1080/09733159.2016.1175131

Steffen, W. et al. (2015). Planetary boundaries: Guiding human development on a changing planet. *Science*, 347. doi:10.1126/science.1259855

Strand, R., Spaapen, J., Bauer, M. W., Hogan, E., Revuelta,G., Stagl, S., Paula, L., and Pereira, A. G. (2015). *Indicators for Promoting and Monitoring Responsible Research and Innovation*. Report from the Expert Group on Policy Indicators for Responsible Research and Innovation. Directorate-General for Research and Innovation, Science with and for Society. Brussels: European Commission

Stern, N.H. (2007). *The Economics of Climate Change: The Stern Review*. Cambridge, UK: Cambridge University Press.

Teräs, J., Salenius, V., Fagerlund, L., and Stanionyte, L. (2018) *Smart Specialisation in Sparsely Populated European Arctic Regions*. EUR 29503 EN. Luxembourg: Publications Office of the European Union. doi:10.2760/960929, JRC114273

Verk, N., Golob, U., and Podnar, K. (2019). A dynamic review of the emergence of corporate social responsibility communication. *Journal of Business Ethics*, 1–25. doi.org/10.1007/s10551-019-04232-6

Webb, M.A. (2014). When and how to pursue corporate social responsibility with core competencies. *Business and Society Review*, 119, 417–434. doi:10.1111/basr.12039

World Economic Forum (WEF). (2020). *The Global Risks Report 2020*. Insight report (15th ed.). Switzerland: Author.

World Wildlife Fund for Nature (WWF). (2016). *Living Planet Report*. Accessed on 11 January 2021 from www.wwf.org.uk/sites/default/files/201610/LPR_2016_summary_spread%20low%20res.pdf

8 CSR and the new trends in business practices and sustainable enterprise

Learning objectives

- Understand how socio-environmental challenges are influencing contemporary CSR trends.
- Understand the mechanisms of the Arctic responsible supply chains.
- Recognize the benefits of aligning with the UN Global Compact (UNGC) and CSR Compass.
- Identify international frameworks (UN Declaration on the Rights of Indigenous Peoples, European Union (EU) Strategy for CSR as part of the EU's 2030 strategy on smart, sustainable, and inclusive growth, and the new integrated EU policy for the Arctic (OECD).
- Understand the circular economy as a model of Arctic development.

Introduction

If the Arctic is to be developed in a sustainable manner, it must attract responsible companies that will play a key role in sustainable economic growth. It could pave the way for a unique approach towards conducting sustainable business in the Arctic. Responsible business practices entail respect for human rights, labour- and environmental standards (including climate and biodiversity), and anti-corruption.

Corporate social responsibility (CSR) in the Arctic means the continuous voluntary commitment of companies to contribute to sustainable development through its business practices. These practices are aligned with ethical principles, respect for cultural differences, respect for intergenerational rights to a healthy environment, capacity building to promote sustainable livelihoods, skills enhancement, and quality of life. CSR in the Arctic consists in integrating social and environmental concerns in business operations and in the interaction with multiple stakeholders.

The existing frameworks and guidelines on CSR for the Arctic are the OECD guidelines for multinational enterprises, the UN Global Compact, the UN Declaration on the Rights of Indigenous Peoples (ILO Convention #169), and the CSR Compass (Nordic Council of Ministers) for Small and

Medium Enterprises. These frameworks enable businesses operating in or with the intention to operate in the Arctic to assess, evaluate, and report their economic, environmental, social, and governance performances.

Types of discussion and issues explored in Chapter 8

Chapter 8 aims to highlight what CSR is for in the Arctic region and provides information on how companies and individuals could both contribute to and increase awareness of what CSR means in the Arctic context. It is correct to say that CSR is a multidimensional concept that cuts across many different areas. Other familiar terms are corporate responsibility, social responsibility, corporate sustainability, and responsible business conduct. This chapter also highlights the most important global frameworks and policies in place and in need of practical implementation.

Arctic responsible supply chains

The influence of the OECD Guidelines for Multinational Enterprises

The complex nature of international business operations in the Arctic and the rapid inflow of foreign direct investment in the region have triggered important reflections on the structure of multinational enterprises, technological innovation, and the strict alignment of business practices with international socio-environmental standards. Good corporate governance principles and practices, in this context, entail good governance of the Arctic region.

The Arctic Council urges companies operating in the Arctic or with the intention to develop Arctic operations to adopt and follow the OECD Guidelines for Multinational Enterprises, part of the OECD Declaration on International Investment and Multinational Enterprises (Declaration on International Investment and Multinational Enterprises, 2011, p. 7). The OECD guidelines consist of

> recommendations addressed by governments to multinational enterprises aiming at ensuring that the operations of these enterprises are in harmony with government policies, to strengthen the basis of mutual confidence between enterprises and the societies in which they operate, to help improve the foreign investment climate and to enhance the contribution to sustainable development made by multinational enterprises.
>
> (OECD/LEGAL/0144)

The Guidelines represent the first international corporate responsibility instrument to incorporate risk-based due diligence into major areas of business ethics (OECD, 2014). The nature and extent of due diligence depend on the specific circumstances of industries, businesses, and geographical contexts in relation to the complexity of international business and supply chains. Despite representing a voluntary framework of principles and standards, the nations that adhere to these parameters are bound to applying and implementing them according to national and international law.

Responsibility must be at the centre of business and industrial operations in the Arctic. Aligned to this principle is the sense of responsible business conduct based on rules, policies, and standards, and based on the interfaces between the UN Guiding Principles on Business and Human Rights as well as the ILO Tripartite Declaration of Principles Concerning Multinational Enterprises and Social Policy (ILO, 1998, 2017; UN, 2011). In March 2019, a working group named 'Responsible Business Conduct' was created by the OECD within the OECD Guidelines for Multinational Enterprises as part of the OECD Declaration on International Investment and Multinational Enterprises (OECD, 2019) (OECD/LEGAL/0144) in order to pursue convergence regarding what are considered 'responsible business practices' according to common standards and guidelines for multinational enterprises.

Areas of priority for the Arctic in terms of sustainable business practices would consist in addressing risks and opportunities in Arctic value chains as well as applying risk management tools to identify risky supply-chain segments. Due diligence efforts oriented to assess the alignment of industry operations and practices to OECD certification programmes as well as the measurement of impacts provided by this alignment are important components of the due diligence process related to CSR practices in the Arctic. Apart from these components, the 'OECD Due Diligence Guidance for Responsible Business Conduct' (OECD, 2014) also focuses on the level of business engagement with stakeholders representing an important tool of convergence and development.

The due diligence system applied to supply chains has the role of identifying and assessing risk in specific strategic supply chains or segments of supply chains in order to establish robust corporate management systems, provide adequate response to risks, and provide adequate auditing and reporting. There are several benefits of implementing risk-based due diligence associated with efficiency gains, licensing granting (social and environmental), diversification, reputation management, and access to markets. The trend is to integrate the domestic and international policy levels of RBC, by reinforcing local and global corporate commitments with practical due diligence instruments. This mainly applies to specific sectors and supply chains requiring policy coherence and related to minerals, agriculture, extractive industry, and finance.

Risk in this context is not only the risk 'interna corporis' related to the enterprise (i.e. operational risk, financial risk, market risk, and reputational risk), but also risk 'externa corporis' associated with adverse impacts of its operations, supply chains, or business relationships on people, the environment, and society. These categories of risks can be identified through inconsistencies between corporate practices and OECD guidelines. Inconsistencies and divergencies in relation to consolidated international standards are probably the most relevant aspect addressed by due diligence instruments in Arctic companies' operations. This is because risks related to adverse impacts can take place in the present and in the future, and the function of the due diligence process is to guarantee alignment between operations and responsible business conduct in the very sensitive region of the Arctic.

The risks at stake are of different orders and can materialize into human rights violations, employment and industrial relations divergences, environmental risks, divergences in consumer interests, disclosure issues, and bribery circumstances.

In the South, companies have been dealing with risks and adverse socio-environmental impacts for a long time. If these risks have serious consequences in corporate operations in the southern operations, they can have even more complex effects in the North. The triple bottom line (Elkington, 1998) generally advocates three orders of impact related to economic, social, and environmental aspects of corporate activities. However, in the Arctic, the triple bottom line is not limited to the traditional three pillars of sustainability because the 'cultural component' makes a huge difference in the sustainable development equation despite still not being considered as a pillar of the triple bottom line (Arruda, 2019, p. 65). Cultural characteristics of models of development (Arruda, 2018, p. 112) have a crucial influence on Arctic business strategies, leadership styles, business models, responsible businesses conduct, and standards.

Considering that there are inherently risky industrial activities and other operations that despite not being considered inherently risky may also present adverse impacts according to specific circumstances, the due diligence systemic process only offers advantages in terms of risk assessment and response in the sense that it assists decision-makers in identifying, assessing, preventing, and mitigating operations, and reviewing business relationships. The process not only avoids and decreases exposure to systemic risks (see Tables 8.1–3) but also optimizes costs, market strategies, stakeholders' relationships, and management systems by adding more value to supply chains and contributing effectively to society.

Another advantage of the due diligence process is the versatility and adaptability to the size of the enterprise, to local and regional characteristics of different jurisdictions (regulation, levels of development), business models, position in the supply chain, and levels of engagement with stakeholders. The latter factor is particularly sensitive when the enterprise has the potential to cause or has caused adverse impacts on people, the economy, environment, or cultural assets, making meaningful stakeholder engagement and trust building relevant components throughout the due diligence process. In Arctic industrial operations, engagement with stakeholders involves non-indigenous and indigenous peoples, which makes the corporate responsibility code of conduct even more complex to design and vital to implement and communicate. In this sense, transparency, trust building, and communication processes are fundamental to encourage and perform co-management (Arruda, 2018) through

Table 8.1 Human rights and environmental risks

Human rights	*Environment*
• Forced labour. • Wage discrimination for equal work or work of equal value. • Gender-based violence or harassment including sexual harassment. • Failing to identify and appropriately engage with indigenous peoples where they are present and potentially impacted by the enterprise's activities. • Involvement in reprisals against civil society and human rights defenders who document, speak out about, or otherwise raise potential and actual human rights impacts associated with projects. • Restriction on people's access to clean water.	• Ecosystem degradation through land degradation, water resource depletion, and/or destruction of pristine forests and biodiversity. • Unsafe levels of biological, chemical, or physical hazards in products or services. • Water pollution (e.g. through discharging waste water without regard to adequate wastewater infrastructure).

Source: Adapted from *OECD Due Diligence Guidance for Responsible Business Conduct* (OECD, 2018)

participative and inclusive instances of agreed decision-making, cooperation, and minimization of risks (social, economic, environmental, and cultural).

Since the benefits of applying due diligence to Arctic enterprises have already been highlighted, it is also important to understand the practical steps to be considered to create Arctic responsible supply chains. This systemic view of the transformation of Arctic management systems towards responsible business practices and conduct entails identifying and assessing impacts; preventing, mitigating, or ceasing these impacts; auditing and tracking implementation plans on mitigation and remediation; and widely communicating how the company addressed these effects. In order to avoid a fragmented process, it is vital to embed this systematic view in the company's management systems by redesigning policies on specific serious risks; aligning standards and policies across teams and business units; amplifying channels of communication and recording systems to disseminate and keep information on risks and policies across the units; providing training; implementing policies and assigning responsibilities of due diligence to managers whose decision-making can influence the degrees of corporate risk. This will probably inspire the creation of an Arctic certification in the near future.

Local governments, as stakeholders, also have their part in the process of promoting implementation of responsible business conduct and practices in the Arctic through the alignment and coherence of policies and regulation to the OECD guidelines. Through governmental collaboration and partnerships with businesses, governments can provide effective domestic

Table 8.2 Consumer interests, employment, and industrial relations

Consumer interests	*Employment and industrial relations*
• Goods and services do not meet all agreed or legally required standards for consumer health and safety, including those pertaining to health warnings and safety information. (OECD, 2011, Chapter VIII, Para 1). • Failing to provide accurate, verifiable, and clear information that is sufficient to enable consumers to make informed decisions, including information on the prices and, where appropriate, content, safe use, environmental attributes, maintenance, storage, and disposal of goods and services. (OECD, 2011, Chapter VIII, Para. 2) • Representations or omissions of information, or any other practices, that are deceptive, misleading, fraudulent, or unfair. (OECD, 2011, Chapter VIII, Para. 4)	• Failing to respect the right of workers to establish or join trade unions or representative organizations of their own choosing and have trade unions and representative organizations of their own choosing recognized for the purpose of collective bargaining. (OECD, 2011, Chapter V, Para. 1.b) • Failing to engage in constructive negotiations, either individually or through employers' associations, with such representatives with a view to reaching agreements on terms and conditions of employment. (OECD, 2011, Chapter V, Para. 1.b) • Child labour, including worst forms of child labour. (OECD, 2011, Chapter V, Para. 1.c) • Discrimination against workers with respect to employment or occupation on such grounds as race, colour, sex, religion, political opinion, national extraction, or social origin, or other status. (OECD, 2011, Chapter V, Para. 1.e) • Failing to adapt machinery, equipment, working time, organization of work, and work processes to the physical and mental capacities of workers. (OECD, 2011, Chapter V, 1.e; ILO, 1981, No. 155) • Failing to replace hazardous substances with harmless or less hazardous substances wherever possible. (ILO, 1988, No.167) • Payment of wages that do not meet the basic needs of workers and their families. (OECD, 2011, Chapter V, Para. 4.b) • Threatening to transfer the whole or part of an operation unit in order to hinder workers from forming or joining a trade union. (OECD, 2011, Chapter V, Para. 7)

Source: Adapted from *OECD Due Diligence Guidance for Responsible Business Conduct* (OECD, 2018)

Table 8.3 Disclosure and bribery

Disclosure	*Bribery*
• Failing to disclose material information on the financial and operating results of the enterprise, enterprise objectives, major share ownership and voting rights, remuneration policy for members of the board and key executives, information about board members, related party transactions, foreseeable risk factors, issues regarding workers and other relevant stakeholders, governance structures and policies. • Failing to provide the public and workers with adequate, measurable, and verifiable (where applicable) and timely information on the potential environment health and safety impacts of the activities of the enterprise.	• Bribery of public officials to win public procurement contracts. • Bribery of public officials to obtain favourable tax treatment or other preferential treatment or access to confidential information. • Bribery of public officials to obtain customs clearance. • Bribery of public officials to obtain authorizations or permits. • Selling products to government agencies at an elevated price to provide public officials with a share of the profit. • Bribing public officials to ignore or avoid regulations or controls. • Providing gifts, meals, and entertainment to those with whom the enterprise does business without adequate controls or records. • Receiving gifts from business partners or public officials without adequate controls or record.

Source: Adapted from *OECD Due Diligence Guidance for Responsible Business Conduct* (OECD, 2018)

policy frameworks aligning with stable macroeconomic policy, adequate and balanced regulation, an impartial law enforcement and judicial system, as well as an efficient and ethical public administration. By adhering to international standards and ethical guidelines, governments can support enterprises in strengthening sustainable development, living standards, and the well-being of society.

Finally, it is important to remark that responsible business practices and conduct in the Arctic is not only about compliance to regulation, international standards, and technological solutions, but also an intrinsic awareness and recognition that business operations and supply chains perform under the integral lenses of the SDGs and Agenda 2030 as the ultimate base to shape and share sustainability policies. Linking the SDGs to companies' performance can represent a step forward in this process of integrating/embedding the SDGs into modern business models of the 21st century. Implementing responsible business conduct through a due diligence process and meaningful engagement with stakeholders can help companies operationalize the SDGs and ensure the evolution of Arctic responsible co-management systems. This means transforming the old southern CSR patterns and debate about risk management to another level, conceptually and practically.

UN Global Compact and CSR Compass: A coherent framework and tool for sustainable business

The mission of the UN Global Compact (UNGC) consists in mobilizing companies and stakeholders according to a vision of alignment of business strategies and operations with principles of human rights, labour, environment, and anti-corruption by synergistically orienting strategic actions to the SDGs. The focus is doing business responsibly through multi-stakeholder initiatives that advance corporate sustainability. The UNGC advocates that:

> Corporate sustainability is imperative for business today – essential to long-term corporate success and for ensuring that markets deliver value across society. To be sustainable, companies must do five things: Foremost, they must operate responsibly in alignment with universal principles and take actions that support the society around them. Then, to push sustainability deep into the corporate DNA, companies must commit at the highest level, report annually on their efforts, and engage locally where they have a presence.
>
> (UNGC, 2014)

In other words, sustainable companies are principled, strengthen society through local action, and instigate leadership and transparent reporting, according to a long-term strategy aligned with financial, social, environmental, and ethical values. However, these are challenging vectors for businesses considering the complexity of Arctic operations because Arctic industrial operations in the face of climate change require a new era of business leadership. A new style of transformative leader is needed who is ready to create solutions for local and global needs. It is an era of action where the strength of global partnerships will be the real game-changer towards a low-carbon society and global well-being.

The UNGC has an important role to play in shaping and sustaining the Arctic responsible supply chains. Through four basic principles of Responsible Business & Leadership Practices, SDGs, Global Platform & Connector, Impact & Performance, the UNGC is gearing its guidelines towards responsible business practices in order to provide a real alignment of business strategies to the UN values and the 2030 Agenda, by mobilizing the global business community into a synergic action oriented by trust, communication, and impact.

The positive aspect of the UNGC is the fact that the advocated principles are the base for engaging businesses, regardless of size, geography, and sector, to align their operations with SDGs. By defining clear objectives and targets for global society to reach by 2030, the SDGs represent an invaluable compass for business planning and new business models to tackle local and global challenges. The collaborative action stimulated by the principles helps companies to share best and emerging practices and to forge partnerships that promote compliance to technical standards, policies, and regulation.

A sustainable company operates its business activities with a genuine concern for not generating negative impacts (externalities) for people, society, the environment, and the economy. It undergoes a constant process of adaptation and innovation, creating models that incorporate long-term strategies for investment and decision-making aligned with the Agenda 2030 and the Theory of Change as a guiding methodology (UNGC, 2019). Its principles and values are an integral part of the day-to-day operations and organizational culture.

The CSR Compass is a tool to support small and medium-sized companies to manage supply chains responsibly in the Arctic. Aligned with national and international best-practice standards, the CSR Compass provides guidelines for companies to design their CSR policies, codes of conduct, checklists, and action plans to implement principles of responsible supply chain management systems.

The CSR Compass represents an additional tool to encourage a process of harmonization, enabling suppliers to design coherent policies and practices, comply with CSR principles, and meet the requirements of businesses and customers. There are several advantages of this alignment, including the assurance of applying equal terms and standards of responsible supply chain management locally and globally, minimization of risks, creation of better opportunities, knowledge sharing, image and reputation, and cooperation.

According to the CSR Compass (2020), responsible supply chain management can generate value for companies based on several factors (see Figure 8.1).

Arctic responsible supply chains require responsible management consisting of coordination and collaboration with supply-chain actors to ensure that CSR principles and strategies are in place to face operational risks, specific industry risks, and local regulation or standards in terms of human rights, labour rights, environmental impacts, and anti-corruption procedures.

UN Declaration on the Rights of Indigenous Peoples

The Arctic has undergone an intense transition over the last ten years and is set to become the main global resource base and a major trade corridor for the next 30 years. The benefits of the Arctic emerging economy may be seen in the creation of local economic development and in meeting the global demand for minerals. Apart from the oil and mining industries, other business opportunities in the Arctic are associated with extraction industries like wood, the food industry (aquaculture, fishing), and tourism. Despite differences in terms of infrastructure, transportation systems, and local constraints to business operations, there are highly competitive companies operating in high-tech industrial environments and performing environmentally friendly activities.

A fundamental particularity of the Arctic economy and business operations is the interaction with indigenous populations. This offers a unique context involving nature, climatic conditions, technology, markets, and new ways of

Figure 8.1 Responsible supply chain management benefits

Figure 8.1 explains the benefits of a responsible supply chain management system. It draws attention to compliance with customers' expectations and requirements, collaboration with suppliers, reduction of the risk of negative PR, improvement of product quality, the creation of new market opportunities, security for investors and creditors, and employee retention.

Source: Adapted from CSR Compact (2020)

doing business associated with Arctic values, culture, and traditions. In this sense, to do business in the Arctic, it is essential to understand the complexity, scale, and specific characteristics of indigenous groups; their historical backgrounds; economic and cultural activities; and the political and regulatory environments.

It is estimated that Indigenous Peoples occupy approximately 20% of the world's land surface, and steward 80% of the planet's biodiversity (Sobrevila, 2008). The indigenous people's attachment to the land is an essential part of their cultural habits. State rules contradicting their rights compromise their intergenerational knowledge and values transmission, and consequently their representativeness (Arruda and Krutkowski, 2016). A sense of place has been identified as an important determinant of physical, mental, and emotional health according to medical and health geographies. This strong sense of connection to place, the environment, and to the organisms that inhabit it is a source of empowerment due to the insightful peculiarities of Traditional Ecological Knowledge, but it can also make them vulnerable to the negative impacts of business development in their ancestral lands and territories. Apart from the physical impacts to 'place', Willox et al. (2013, p. 539) argue that industrial activities can produce severe disruptions to socio-cultural land activities (hunting, fishing, reindeer herding, foraging, as well as travelling on snow and ice); changes to individual and collective senses of place attachment, as well as changes in the mental conceptions of place.

Another aspect to consider in terms of the interaction between indigenous peoples and businesses is the sense of distrust in companies based on historical experiences of economic exploitation, irremediable land degradation (oil and gas industry), human rights abuses, lack of legal protection, marginalization, poor health conditions, poverty, and the lack of understanding of consultative procedures. Physical alterations are closely followed by socio-cultural changes – due to the influx of a transient workforce from outside who bring with them different lifestyles and gender imbalances, causing the Arctic to become an unfamiliar culturescape (Arruda and Krutkowski, 2017).

Climate change and its permafrost transformations, the increased industrialization of the Arctic, and new Arctic trade routes have gradually resulted in a change from nomadic living to sedentary communities, which in turn have led to changing lifestyles, priorities, and values. A complex process of adaptation is on course, not only involving physical and geographical adaptation but a more in-depth process of multicultural encountering and co-creating of sustainable futures. This is a very delicate balance that offers both challenges and opportunities for businesses when engaging with indigenous peoples in new entrepreneurship ventures. At the same time, it is a moment of mutual learning and multicultural understanding of business practices, purposes, impacts, and benefits for both the business and indigenous communities of the Arctic.

This process of business engagement with Arctic indigenous peoples will be crucial to promote a wide debate about social and economic expectations for both parties, as well as to engage in the debate about the meaning of sustainable development for both groups. This collaborative synergy will only be possible by optimizing ecosystem services based on the recognition of the value of Traditional Knowledge, by optimizing the participation of indigenous groups in socio-environmental risk assessments, and by respecting indigenous culture and rights according to the international legal framework. We believe that once the rights of indigenous peoples are jeopardized, businesses incur significant legal, financial, and reputational risks. Joint assessment and management of operational impacts must be based on mutual consent, information exchange, responsible coordination, and cooperation oriented towards a customized model of resources co-management for specific geographical locations.

European Union (EU) strategy for CSR as part of the EU's 2030 strategy on smart, sustainable, and inclusive growth and the new integrated EU policy for the Arctic

CSR is of paramount importance for the European Union (EU), not only for the EU economy, but for enterprises and society. It is seen as an important mechanism of risk management, access to capital, cost optimization, but also expedient for the harmonic relationships among stakeholders, the sustainability and innovation of business operations, and therefore a sustainable economic system and society.

For the European Commission, CSR means the responsibility of enterprises for their impact on society. Companies exert CSR leadership when, apart from complying with the regulation, they integrate social, environmental, ethical, consumer, and human rights concerns into their business strategy and operations (European Commission, 2020).

The European Commission advocates the position that:

> the actions of companies have significant impacts on the lives of citizens in the EU and around the world, not just in terms of the products and services they offer or the jobs and opportunities they create, but also in terms of working conditions, human rights, health, the environment, innovation, education and training.
>
> (European Commission, 2020)

With this in mind, the EU CSR strategy aims at stimulating enterprises of all sizes to adhere to and align with European and global CSR approaches, principles, and guidelines. Moreover, the EU strategy encourages the sharing of good practices and the integration of CSR into research, education, and training through co-regulation, disclosure, and transparency in terms of socio-environmental information (European Commission, 2020).

In the same spirit of integrated action and cooperation in research, science, innovation, and business according to priority areas and projects aiming to address climate change and implement a sustainable use of resources, the European Commission (2016) and the High Representative for Foreign Affairs and Security Policy elaborated and adopted the so-called 'new integrated EU policy for the Arctic'. The policy represents the EU guidelines for the Arctic region with a focus on promoting sustainable development in a highly globally sensitive and strategic area of socio-environmental, economic, and cultural importance.

The EU guidelines for the Arctic articulated 39 actions focused on three main policy objectives (see Figure 8.2).

The coordination and cooperation between the EU and the Arctic Council enhances coherence between Arctic states' national policies and EU policies towards well-oriented measures of environmental protection (oceans, ecosystems, biodiversity), social respect for livelihoods and cultural assets, climate change impacts, pollutants (POPs and black-carbon sources), energy production, extractive industry, tourism, and maritime shipping. It also enhances understanding of the interdependence between the Arctic and non-Arctic world as well as commitment and compliance with existing EU legislation and the Arctic policies of the EU Member States. It identifies general and specific areas of action towards developing a more coherent framework for EU action with a focus on the European Arctic.

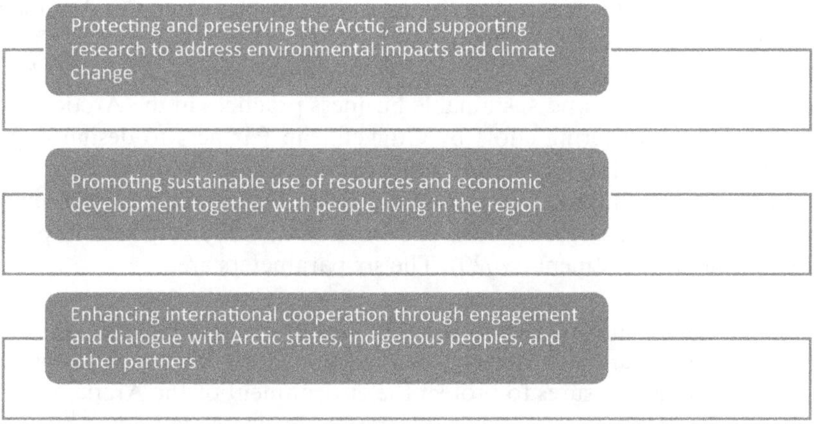

Figure 8.2 EU Arctic policy objective areas

Figure 8.2 is especially important for outlining the main components of Arctic policy in protecting and preserving the Arctic, supporting research to address environmental impacts and climate change, but also enhancing the sustainable use of resources and economic development together with people living in the Arctic. It highlights the importance of enhancing international cooperation through engagement and dialogue with Arctic states, indigenous peoples, and other partners.

Source: Adapted from European Commission (2016)

The EU advocates a balanced combination of Arctic industrial specialization allied to science-informed decision-making with respect to Traditional Knowledge to reduce risks, enable adaptation of local communities, and promote sustainable development according to the parameters established by the SDGs and the Agenda 2030. Joint efforts of extending information analysis and communication, research leadership, ecological innovation, youth employment, and knowledge sharing among collaborators contribute, in practice, to the efficient implementation of regional and international frameworks. In addition, these efforts enable the efficient incorporation of CSR into the activities of companies operating businesses in the Arctic region. Representatives of the business sector (Arctic Economic Council) and companies' application of voluntary mechanisms raise the bar in terms of industrial standards (Arctic Environmental Responsibility Index) for business performances according to the Arctic Investment Protocol and the UN Global Compact Initiative.

Arctic Investment Protocol

The Arctic Investment Protocol from 2016 was a pioneering effort of Guggenheim, a global investment and advisory financial services firm. The aim was to advance responsible and sustainable business practices in the Arctic. The Protocol was then transferred by the World Economic Forum to the Arctic Economic Council (ACE), an independent organization that facilitates Arctic business activities and responsible economic development (Guggenheim Investments, 2020).

The Arctic Investment Protocol is a contemporary vision of how to advance responsible and sustainable business practices in the Arctic. It is the product of a pioneering effort by Guggenheim Partners to design objective criteria (six principles) for promoting equitable and sustainable economic growth in the Arctic towards well-being and resilience for local communities based on fairness, inclusiveness, and environmentally responsible practices (Guggenheim Investments, 2020). The six parameters are:

1. Build resilient societies through economic development.
2. Respect and include local communities and indigenous peoples.
3. Pursue measures to protect the environment of the Arctic.
4. Practice responsible and transparent business methods.
5. Consult and integrate science and traditional ecological knowledge.
6. Strengthen pan-Arctic collaboration and sharing of best practices.

(WEF, n.d.)

Circular economy as a model of Arctic development

Promoting the balance among the pillars of sustainability is challenging but it has never been so fundamental to respond to this pressing contemporary dilemma. In the last decade, there have been more questions than answers because society does not know how to create prosperity for a growing global population and at the same time strengthen the Earth's support systems. This is the current learning for industries, societies, and citizens. There is no plausible answer from governments or businesses on how to achieve continued economic development while preserving the resource base that fuels the global economy. It is not the first time that this idea of rethinking the way global society operates has been debated and repeated, at least over the last two decades, in papers and books.

The current Arctic context of change and the importance of responsible business conduct bring on board the model of the circular economy, which seems to hold the promise of promoting the so-desired balance and reconciliation with these seemingly opposing objectives in order to create long-term value and cross-chain synergy. The trend is being felt in different sectors. This model of taking raw materials (natural resources) and outputting waste to the environment is obsolete and is about to be replaced by a circular, restorative new system based on well-functioning natural systems and a stable and green supply chain capable of minimizing the societal footprint and generating green jobs.

According to a more functional concept and to Bourguignon (2016), "the circular economy is a model of production and consumption, which involves sharing, leasing, reusing, repairing, refurbishing and recycling existing materials and products as long as possible". Based on the Life Cycle Theory, this model reduces waste to a minimum by keeping materials within the economy for longer. This involves extending the life cycle of the product and breaking the linear economic model based on perceived or planned obsolescence and the "take-make-consume-throw away pattern" of production and consumption (EEA, 2017). CSR has a crucial role in shaping a new paradigm for the business of tomorrow as well as in designing practical solutions aligned with the circular economy by creating value and relying on cheap, accessible materials and energy. Several Nordic countries have already developed national strategies (CSR action plans) based on new patterns of production and consumption, which will have expressive impacts on trade, consumers, and on the social and environmental conditions of globalized production.

Further resources

There are several other perspectives that can be approached by the alignments of this chapter when studying CSR and the new trends in business practices and sustainable enterprise. The application of the circular economy as a model of Arctic development together with the SDGs and the Agenda 2030, CSR Global Compass, and the UN Declaration of Indigenous Peoples Rights opens the path for creative business and entrepreneurship.

In 'Motivations for Indigenous Island entrepreneurship: Entrepreneurs and behavioral economics in Greenland', indigenous entrepreneurship seems driven by intertwined motivations and certain communal values, such as supporting the local community, cultural pride, family, and place. The paper contributes to the understanding of entrepreneurship as an instrument for decolonization, drawing on a range of empirical cases and a multi-disciplinary approach (Wennecke et al., 2019).

The progressive and irreversible transition to a sustainable economic system has been revealed to be a great vector of creativity and inclusion. In 'Women entrepreneurs in Northern Canada: Contexts and challenges', a perspective on the contexts of, and challenges facing, female entrepreneurs in Northern Canada is presented. The study draws on previous research, analysis of survey data, and interviews discussing women's Indigenous and non-Indigenous entrepreneurship in Northern Canada (Orser and Riding, 2016).

Questions for discussion

1. How can the circular economy be seen as a model of Arctic development?
2. How are socio-environmental challenges influencing contemporary CSR trends?
3. What are the principles and mechanisms of the Arctic responsible supply chains?
4. What is the importance of international frameworks in shaping smart, sustainable, and inclusive policies for the Arctic and for the world?

Chapter summary

One fundamental characteristic of the Arctic economy and business operations is the interaction with indigenous populations that offers a unique context involving nature, climate change, technology, markets, and new ways of doing business associated with Arctic values, culture, and traditions. The decisions, actions, and the *modus operandi* of businesses and entrepreneurs have significant impacts on the lives of Arctic citizens. These components also impact products and services offered, the jobs and opportunities created, working conditions, human rights, health, environment, innovation, education, and training in the Arctic. A balanced combination of Arctic industrial specialization allied with science-informed decision-making, efficient implementation of regional and international frameworks with respect to Traditional Knowledge, a risk-reduction approach, and adaptation and change co-management with local communities will help to promote sustainable development (SD) according to the parameters established by the SDGs and the Agenda 2030.

References

Arruda, G.M. (2018). Arctic resource development: A sustainable prosperity project of co-management. In G.M. Arruda (Ed.), *Renewable Energy for the Arctic: New Perspectives* (pp. 106–117). Abingdon, UK: Routledge.

Arruda, G.M. (2019). *Sustainable Energy Education in the Arctic: The Role of Higher Education* (1st ed.). Abingdon, UK: Routledge.

Arruda, G. and Krutkowski, S. (2016). Arctic governance, indigenous knowledge, science and technology in times of climate change: Self-realization, recognition, representativeness. *Journal of Enterprise Communities: People and Places in the Global Economy*, 11(4), 514–528.

Arruda, G.M. and Krutkowski, S. (2017). Social impacts of climate change and resource development in the Arctic: Implications for Arctic governance. *Journal of Enterprising Communities: People and Places in the Global Economy*, 11(2), 277–288.

CSR Compass. (2020). Benefits of responsible supply chain management. Accessed on 28 June 2020 from www.csrcompass.com/benefits-responsible-supply-chain-management-0

Declaration on International Investment and Multinational Enterprises. (May 2011). Annex 1. Guidelines for Multinational Enterprises: Recommendations for Responsible Business Conduct in a Global Context. Accessed on 20 January 2021 from https://legalinstruments.oecd.org/public/doc/241/body-text.en.html; www.oecd.org/daf/inv/investment-policy/ConsolidatedDeclarationTexts.pdf

Elkington, J. (1998). Accounting for the triple bottom line. *Measuring Business Excellence*, 2(3), 18–22. https://doi.org/10.1108/eb025539.

European Commission. (2016). An integrated EU policy for the Arctic. Accessed on 28 June 2020 from https://ec.europa.eu/environment/efe/news/integrated-eu-policy-arctic-2016-12-08_en#:~:text=8%20December%202016-,An%20integrated%20EU%20policy%20for%20the%20Arctic,enormous%20environmental%20and%20economic%20importance

European Commission. (2020). Corporate social responsibility & responsible business conduct. Internal market, industry, entrepreneurship and SMEs. Accessed on 28 June 2020 from https://ec.europa.eu/growth/industry/sustainability/corporate-social-responsibility_en

European Environment Agency (EEA). (2017). *Circular by Design: Products in the Circular Economy*. EEA Report No 6/2017. Brussels: Author.

International Labour Organization (ILO). (1998). ILO Declaration on Fundamental Principles and Rights at Work. Accessed on 15 December 2020 from www.ilo.org/declaration/lang--en/index.htm

International Labour Organization (ILO). (2017). ILO Tripartite Declaration of Principles concerning Multinational Enterprises and Social Policy, 5th Edition. Accessed on 15 December 2020 from www.ilo.org/empent/areas/mne-declaration/lang--en/index.htm

Organisation for Economic Co-operation and Development (OECD) (2011). *OECD Guidelines for Multinational Enterprises*. Accessed on 31 May 2020 from https://mneguidelines.oecd.org/guidelines/.

Organisation for Economic Co-operation and Development (OECD). (2014). *OECD Guidelines for Multinational Enterprises: Responsible Business Conduct Matters*. Accessed on 31 May 2020 from https://mneguidelines.oecd.org/MNEguidelines_RBCmatters.pdf

Organisation for Economic Co-operation and Development (OECD). (2014). *Annual Report on the OECD Guidelines for Multinational Enterprises 2014: Responsible Business Conduct by Sector*. Paris: OECD Publishing. https://doi.org/10.1787/mne-2014-en

Organisation for Economic Co-operation and Development (OECD). (2018). *OECD Due Diligence Guidance for Responsible Business Conduct*.

Organisation for Economic Co-operation and Developmen (OECD). (2019). *Annual Report on the OECD Guidelines for Multinational Enterprises 2018*. OECD/LEGAL/0144.

Orser, B. and Riding, A. (2016). Women entrepreneurs in Northern Canada: Contexts and challenges. *International Journal of Entrepreneurship and Small Business*, 27(2/3).

Sobrevila, C. (2008). *The Role of Indigenous Peoples in Biodiversity Conservation: The Natural but Often Forgotten Partners*. Washington, D.C.: The International Bank for Reconstruction and Development.

United Nations (UN). (2011). UN Guiding Principles on Business and Human Rights. Accessed on 15 December 2020 from https://business-humanrights.org/en/un-guiding-principles

United Nations Global Compact (UNGC). (2014). *Guide to Corporate Sustainability: Shaping a Sustainable Future*. New York: Author.

United Nations Global Compact (UNGC). (2019). *UN Global Compact 2019: Annual Management Report*. New York: UNGC Board, March 2020. Author.

Wennecke, C.W., Jacobsen, R.B., and Ren, C. (2019). Motivations for Indigenous island entrepreneurship: Entrepreneurs and behavioral economics in Greenland. *Island Studies Journal*, 14(2), 43–60.

9 CSR auditing and reporting

Learning objectives

- Understand auditing and reporting in an Arctic context.
- Understand that the rights of Indiegenous Peoples must be accounted for when reporting on Arctic operations.
- Understand that enterprises must consider various tools, principles, conventions, guidelines, and standards when reporting on Arctic issues.
- Evaluate whether the Danish CSR compass is useful for selecting appropriate tools, principles, conventions, guidelines, and standards for reporting on Arctic issues.

Introduction

Corporate social responsibility (CSR) auditing and reporting address the CSR performance of businesses. This means for instance ensuring that they are respecting human and labour rights, taking into account consumer and other stakeholders' interests, minimizing negative environmental impacts, or preferably compensating for environmental damage, and preventing corruption. CSR performance thus signals to what degree companies comply with CSR principles, with auditing serves as a way of verifying companies' reports. To ensure corporate compliance with Arctic-specific CSR guidelines, corporations have a duty to undertake reporting on CSR practices and policies, meaning obligations to comply with internationally recognized CSR principles, as well as their respective implementation processes. Transparency and dissemination are fundamental factors to ensure CSR impacts in the Arctic region.

The Arctic Council's commitment to CSR is emphasized in the 2013 Kiruna Arctic Council Ministerial Declaration. The Declaration states that the Ministers:

> Welcome the Arctic Council's work on corporate social responsibility and sustainable business, and encourage enterprises operating in the Arctic to respect international guidelines and principles. By this the Arctic Council

> Recognize[s] the central role of business in the development of the Arctic, and decide[s] to increase cooperation and interaction with the business community to advance sustainable development in the Arctic.

To frame what Arctic auditing and reporting is about, the Arctic Council further states that businesses need to behave ethically and should "contribute to economic development while improving the quality of life of the workforce, their families, the local community and society at large". In addition, CSR should be about building capacity for sustainable livelihoods and take into account cultural differences (Arruda, 2019), building skills of employees and the community, while preserving the natural environment for future generations. CSR as a voluntary contribution of businesses should contribute to a better society and unpolluted environment and these concerns need to be integrated into core business operations through interactions with stakeholders. Instead of developing Arctic-specific CSR guidelines and principles, the Arctic Council refers to international guidelines and principles, namely the Global Reporting Initiative, the OECD Guidelines for Multinational Enterprises, and the UN Global Compact (SDWG, 2013), all of which are addressed in this chapter.

Types of principles, guidelines, and standards addressed in Chapter 9

This chapter will focus on the auditing and reporting tools, as well as conventions, principles, guidelines, and standards that are recognized as fundamental in the Arctic context, and thus should be taken into account when reporting on Arctic operations. These are the UN Global Compact, the Global Reporting Initiative, the OECD Guidelines for Multinational Enterprises, the Sustainability Accounting Standards Board, the International Integrated Reporting Council, AccountAbility, Social Accountability 8000, Nasdaq ESG Reporting Guide, the International Labour Organization (ILO) conventions, the UN Declaration on the Rights of Indigenous Peoples, the UN Principles on Businesses and Human Rights, the International Organization for Standardization Standards, the Carbon Disclosure Project, UN Principles for Responsible Investment, and the Arctic Investment Protocol. Some other tools and guidelines are mentioned, mainly guidelines devolved by the European Commission, the European Eco-Management and Audit Scheme, and a good practice guide from the International Council on Mining and Metals. The case studied in Chapter 9 is the Danish CSR Compass, a tool developed to guide companies in the right direction by selecting national and international tools useful to companies in their CSR work, including auditing and reporting.

Background: From voluntary to mandatory CSR reporting

Although CSR is usually referred to as voluntary, some aspects are becoming mandatory, such as CSR reporting of large companies. For instance, EU

Directive 2014/95 now requires large companies to disclose non-financial and diversity information. This directive offers an amendment to Directive 2013/34/EU. Companies that need to comply with the rules from 2018 and onward are large public-interest companies with 500 employees or more. Public-interest companies are listed companies, banks, insurance companies, and other companies designated by national authorities as public-interest entities. Companies' policies need to take into account: a) environmental protection, b) social responsibility and treatment of employees, c) respect for human rights, d) anti-corruption and actions against bribery, and e) diversity on company boards in terms of age, gender, educational and professional background of board members. Although it is mandatory for companies to report, there is still flexibility regarding how and what they report. It is however emphasized that they use international, European, or national reporting guidelines, such as the US Global Compact, the OECD Guidelines for Multinational Enterprises, and/or the International Organization for Standardization – ISO 26000 guidance on social responsibility (European Commission, n.d.a). The European Commission also offers non-binding guidelines on non-financial reporting (European Commission, 2017) and guidelines on reporting climate-related information which supplement reporting climate-related information. The EU Directive 2014/95 is relevant in the Arctic region given that Denmark, Finland, and Sweden are European Union member states, and Iceland and Norway are part of the European Economic Area (EEA) and the EEA Treaty on commerce, meaning that Iceland, Norway, and Liechtenstein (also an EEA member) have agreed to enact similar legislation to that passed in the EU in areas including consumer protection, company law, the environment, and social policy.

To provide an example of how EU Directive 2014/95 was integrated into Icelandic legislation, it was done by making amendments to Law no. 3/2006 on annual reports. In this case, the law applies to companies with more than 250 employees and/or of particular financial strength (article 2.11 c and d). In a similar manner as in the EU Directive they need provide information, "necessary to assess the company's development, scope, status and impact in relation to environmental, social and human resources issues" (article 66.d). The company should furthermore have a policy addressing human rights issues and how the company responds to corruption and bribery issues (article 66.d).

This is however not the only pressure on companies to issue CSR reports. The pressure also comes from the investment community, which frames CSR topics under the heading of environmental, social and governance (ESG) investing. In order to evaluate investment options, CSR, sustainability, or intergraded reports are emphasized, as well as Principles for Responsible Investment (PRI) (Doni and Johannsdottir, 2019). By systematically integrating ESG aspects into the core business of companies, they are able to identify and manage ESG opportunities and risks, recognizing which issues are of strategic importance. They can embed ESG aspects into companies' strategies and culture in order to drive businesses' long-term performance (KPMG, n.d.). By doing so they will also minimize the risk of institutional investors, while increasing the likelihood of positive financial yields.

The UN Global Compact

The UN Global Compact is the largest corporate sustainability initiative. This non-binding United Nations pact is based on commitments of CEOs to ten Principles that are universally acknowledged. By doing so, they are also demonstrating support for achieving the Sustainable Development Goals (SDGs) by 2030. The UN Global Compact was announced by then UN Secretary-General Kofi Annan at the World Economic Forum in 1999, and officially launched at the UN Headquarters in New York City in 2000. The UN Global Compact Principles derive from the Universal Declaration of Human Rights, the International Labour Organization's Declaration on Fundamental Principles and Rights at Work, the Rio Declaration on Environment and Development, and the United Nations Convention Against Corruption (UN Global Compact, n.d.). The Ten Principles are in four categories: human rights, labour, environment, and anti-corruption; see Table 9.1.

The Principles are very much linked to corporate sustainability, integrated through CSR strategies, policies, and procedures. Companies' approach to doing business is grounded in the principles and values shown in Table 9.1.

Table 9.1 UN Global Compact Ten Principles

Human rights	Principle 1: Businesses should support and respect the protection of internationally proclaimed human rights; and Principle 2: make sure that they are not complicit in human rights abuses.
Labour	Principle 3: Businesses should uphold the freedom of association and the effective recognition of the right to collective bargaining; Principle 4: the elimination of all forms of forced and compulsory labour; Principle 5: the effective abolition of child labour; and Principle 6: the elimination of discrimination in respect of employment and occupation.
Environment	Principle 7: Businesses should support a precautionary approach to environmental challenges; Principle 8: undertake initiatives to promote greater environmental responsibility; and Principle 9: encourage the development and diffusion of environmentally friendly technologies.
Anti-Corruption	Principle 10: Businesses should work against corruption in all its forms, including extortion and bribery.

Source: UN Global Compact (n.d.)

In acting in a reasonable manner towards people and planet, companies are also safeguarding their long-term future success (UN Global Compact, n.d.).

The UN Global Compact offers further guidance by identifying five defining features of corporate sustainability (United Nations Global Compact, 2014):

1. **Principled business:** Aligning with the Ten Principles on human rights, labour, environment, and anti-corruption.
2. **Strengthen society:** Taking action and collaborating with others to advance global challenges.
3. **Leadership committment:** Effecting long-term change begins with a company's leadership.
4. **Reporting progress:** Transparency in business practice is crucial for sustainability.
5. **Local action:** Viewing sustainability through a local lens.

Additionally, a framework implementation of the Ten Principles is presented in a UN Global Compact Management Model, including six important elements: 1) commit, 2) assess, 3) define, 4) implement, 5) measure, and 6) communicate. This should be considered as a cycle of continuous improvements (UN Global Compact and Deloitte, 2010).

The Global Reporting Initiative

The Global Reporting Initiative (GRI) is an independent international organization leading in sustainability reporting since 1997. It offers widely adopted GRI Standards for sustainability reporting, in particular by the world's largest corporations. The purpose of GRI is to inspire accountability by helping organizations seize new opportunities, and to identify and manage risks. The focus areas of GRI in the coming years are to (GRI, n.d.a):

1. Create standards and guidance to advance sustainable development.
2. Harmonize the sustainability landscape.
3. Lead efficient and effective sustainability reporting.
4. Drive effective use of sustainability information to improve performance.

The GRI Standards superseded previous GRI Guidelines that were phased out in 2018. The first set of Guidelines were issues in 2000 as the first global comprehensive framework for sustainability reporting. The second generation of the Guidelines (G2) was issued in 2003, the third generation (G3) in 2006, and the fourth generation (G4) in 2013. The same year, GRI joined forces with the UN Global Compact and the World Business Council for Sustainable Development (WBCSD) to craft sector-specific guidance that would help companies within each sector to align their management and reporting with global Sustainable Development Goals and targets (GRI, n.d.b). Table 9.2 shows three universal standards: GRI 101 (foundation),

Table 9.2 Overview of the set of GRI Standards and topic-specific disclosures

Universal standards	*Society*
gri-101-foundation-2016	gri-401-employment-2016
gri-102-general-disclosures-2016	gri-402-labor-management-relations-2016
gri-103-management-approach-2016	gri-403-occupational-health-and-safety-2018
Economic	gri-404-training-and-education-2016
gri-201-economic-performance-2016	gri-405-diversity-and-equal-opportunity-2016
gri-202-market-presence-2016	gri-406-non-discrimination-2016
gri-203-indirect-economic-impacts-2016	gri-407-freedom-of-association-and-collective-bargaining-2016
gri-204-procurement-practices-2016	gri-408-child-labor-2016
gri-205-anti-corruption-2016	gri-409-forced-or-compulsory-labor-2016
gri-206-anti-competitive-behavior-2016	gri-410-security-practices-2016
Environment	gri-411-rights-of-indigenous-peoples-2016
gri-301-materials-2016	gri-412-human-rights-assessment-2016
gri-302-energy-2016	gri-413-local-communities-2016
gri-303-water-and-effluents-2018	gri-414-supplier-social-assessment-2016
gri-304-biodiversity-2016	gri-415-public-policy-2016
gri-305-emissions-2016	gri-416-customer-health-and-safety-2016
gri-306-effluents-and-waste-2016	gri-417-marketing-and-labeling-2016
gri-307-environmental-compliance-2016	gri-418-customer-privacy-2016
gri-308-supplier-environmental-assessment-2016	gri-419-socioeconomic-compliance-2016

Source: GRI (n.d.c)

GRI 102 (general disclosure), and GRI 103 (management approach), and three topic-specific standards: GRI 200 (economic), GRI 300 (environment). and GRO 300 (social) (GRI 101: Foundation 2016, 2018b). The colours in Table 9.2 represent those used for each set of GRI standards (GRI, n.d.c)

GRI 411 focuses specifically on the rights of indigenous peoples and as such setting out requirements, or mandatory instructions, on which issues should be addressed and in what way. The standard claims that while there is no universal definition of indigenous peoples, relevant definitions may be found in the International Labour Organization (ILO) Convention 169, *Indigenous and Tribal Peoples Convention*, from 1989 and the United Nations (UN) Declaration, *United Nations Declaration on the Rights of Indigenous Peoples*, from 2007 (GRI, 2018a). The reporting requirements are provided in Table 9.3.

The OECD Guidelines for Multinational Enterprises

The OECD Guidelines for Multinational Enterprises, an annex to the OECD Declaration on International Investment and Multinational Enterprises, contain suggestions from governments to multinational enterprises on how to conduct business in a responsible way. The Guidelines, which are non-legally

Table 9.3 GRI 411 Rights of Indigenous Peoples

Reporting requirements:

1.1 The reporting organization shall report its management approach for the right of indigenous peoples using *GRI 103: Management Approach.*

Disclosure 411-1: Incidents of violations involving rights of indigenous peoples

The reporting organization shall report the following information:

a) Total number of identified incidents of violations involving the rights of indigenous peoples during the reporting period.

b) Status of the incidents and actions taken with reference to the following:

I. Incident reviewed by the organization;

II. Remediation plans being implemented;

III. Remediation plans that have been implemented, with results reviewed through routine internal management review processes;

IV. Incidents no longer subject to action.

Source: GRI (2018, pp. 5 and 7)

binding on companies but binding on signatory governments, cover issues related to human rights, labour rights, and the environment, information disclosure, combating bribery, consumer interests, competition, taxation, and intellectual property rights. Given the recommendation for information disclosure, they are of relevance for this chapter on auditing and reporting (OECD Watch, n.d.). The Guidelines specifically claim that companies may need to consider additional standards, such as individuals belonging to particular groups or populations where adverse human rights impacts may occur, including Indigenous Peoples. In that case, enterprises should take into account United Nations instruments, some of which are covered in this section, but particular note should be made of the ILO Human Resources Development Recommendation 195 of 2004 where enterprises are encouraged to invest in training and lifelong learning of employees, and ensure equal opportunities to training for women and other vulnerable groups including Indigenous Peoples (OECD, 2011).

According to the OECD Guidelines for Multinational Enterprises, if adverse impacts are caused, or potentially caused, by an enterprise, then it should cease the potential impact or remedy the actual impact. If the adverse impact is contributed to by an enterprise, it should cease or prevent its contribution or use leverage, meaning if it has the ability to effect change, to mitigate any remaining impacts to the greatest extent possible. In the case where adverse impact can be directly linked to enterprise operations, products, or services provided by a business relationship, it should use leverage to influence the entity causing the adverse impact to prevent or mitigate the impact (OECD, 2017).

Of specific relevance in the Arctic is the OECD Due Diligence Guidance for Meaningful Stakeholder Engagement in the Extractive Sector for practitioners in the mining, oil, and gas industries. This is a practical framework used to identify and manage risk regarding stakeholder engagement activities. The purpose is to ensure that companies play an active role to avoid and address adverse impacts in accordance with the OECD Guidelines for Multinational Enterprises. The Due Diligence Guidance includes a framework for assessing stakeholder engagement and consultation performance with regard to specific stakeholder groups including Indigenous Peoples (OECD, 2018). Annex B of the Due Diligence Guidance particularly focuses on how to engage with Indigenous Peoples. Extractive industries should therefore be aware of their unique situation. This means that companies must understand the context of their operation, meaning the regulatory context, the area of impact, land rights, self-governance, historical marginalization or discrimination, and Indigenous People's cultural and spiritual heritage. Furthermore, companies should ensure that Indigenous Peoples are appropriately identified and prioritized. Accordingly, necessary support systems for meaningful engagement should be in place, as well as processes for engagement. In addition, best practices ought to be identified and applied (OECD Publishing, 2017).

Sustainability Accounting Standards Board

The Sustainability Accounting Standards Board (SASB) is a non-profit organization that sets financial reporting and sustainability accounting standards that can be used as a part of companies' core reporting system, either as a standalone reporting system or as a part of an integrated reporting system (SASB, 2018a). SASB offers 77 industry-specific standards, based on shared sustainability opportunities and risks, which offer accounting metrics for each disclosure topic, a technical protocol for compiling of data, and activity metrics for normalization of data. The disclosing standards relate to the concept of materiality of sustainability issues that may have significant impact on financial conditions and performance. (SASB, n.d.). Materiality and impact means the "effect an organization has on the economy, the environment and/or society, which in turn can indicate its contribution to sustainable development" (GRI, 2018b, p. 8). A specific SASB Materiality Map® identifies the sustainability issues to be materialized in different industries. In total, 26 sustainability issues are identified, categorized as general issues, environment, social capital, human capital, business models and innovation, and leadership and governance. The industries covered in the SASB Materiality Map® are consumer goods, extractives and mineral processing, financial, food and beverages, health care, infrastructure, renewable resources and alternative energy, resource transformation, technology and communication, and transportation. Each industry is then further subcategorized into sectors such as coal, oil, and gas, exploration and production, metals and mining, just to name a few (SASB, 2018b). Of these, many industries are relevant for sustainable Arctic economic development, including extractives and mineral processing, infrastructure, renewable resources and alternative energy, and transportation.

The International Integrated Reporting Council

The International Integrated Reporting Council (IIRC) is a global coalition of various stakeholders, including regulators, investors, companies, standard setters, the accounting profession, academia, and NGOs, which offers a principles-based non-mandatory Integrated Reporting Framework. The purpose of IIRC's work, and its Framework, is to establish and influence adoption and improved quality of integrated reporting internationally. This is done through webinars, sharing of best practices, and by addressing practical challenges companies are faced with, such as compliance, materiality, and completeness and conciseness of reporting (IIRC, n.d.). The Integrated Reporting Framework can be used for the adoption of integrated reporting. It offers seven Guiding Principles of reporting explaining what is expected in terms of content and information presented (The International Integrated Reporting Council, 2013):

1. Strategic focus and future orientation.
2. Connectivity of information.
3. Stakeholder relationships.
4. Materiality.
5. Conciseness.
6. Reliability and completeness.
7. Consistency and comparability.

Integrated reports should include information about the organization and its external environment, its governance structure, business model, risks and opportunities, strategy and resource allocation, performance, outlook, basis of preparation and presentation, and general reporting guidance. These elements fundamentally link together and are not mutually exclusive (The International Integrated Reporting Council, 2013). The Integrated Reporting Framework is not Arctic specific but offers a way to ensure the quality of integrated reports, and as such is of relevance for companies that want to demonstrate responsibility with regard to sustainability practices.

AccountAbility

AccountAbility, established in 1999, offers the AA1000 Series of principles-based Standards aimed at offering companies a way to demonstrate a "leadership and performance in accountability, responsibility and sustainability" (AccountAbility, 2020). The foundation of the Series of Standards is the AA1000AP (2018) AccountAbility Principles. There are four Accounting Principles (AccountAbility, 2018a):

1. Inclusivity.
2. Materiality.

3. Responsiveness.
4. Impact.

Of these, Principle 4, impact of behaviour, outcomes and performance, is of central importance for the other Principles, serving to interact with them (AccountAbility, 2018a). The AA1000AP (2018) standard has two companion standards: the AA1000AS (2008, with 2018 Addendum) Assurance Standard, and the AA1000SES (2015) Stakeholder Engagement Standard. The former standard of assurance offers a way for aligning financial reporting and assurance with non-financial aspects of sustainability. It is thus of relevance for sustainability assurance practitioners and providers, and also for the experts preparing sustainability reports (AccountAbility, 2018b). The latter serves as a framework for assessing, designing, implementing, and communicating about stakeholder engagement. It is therefore of key importance for Principle 1, inclusivity, meaning "the participation of stakeholders in developing and achieving an accountable and strategic response to sustainability" (AccountAbility, 2018c, p. 4). The standard offers three generations of stakeholder engagement, namely, an "ad-hoc reactive engagement in response to pressure and change" (first generation), "planned and systematic engagement to manage risk and returns" (second generation), and "integrated strategic engagement for transformational and sustainable performance" (third generation) (AccountAbility, 2018c, p. 6). All of these standards are of relevance for Arctic business operations, transparency, and reporting quality, although the AA1000SES (2015) Stakeholder Engagement Standard is particularly important as it should provide locals and indigenous peoples with a voice when communicating with businesses operating in the region.

Social Accountability 8000

Social Accountability 8000, or the SA8000® Standard, is an auditable certification standard aimed at encouraging businesses to develop, maintain, and apply workplace practices that are socially acceptable. It was established in 1997 by Social Accountability International (SAI) as a multi-stakeholder initiative. It measures social accountability performance in nine business areas that align with the Universal Declaration of Human Rights and International Labour Organization (ILO) conventions, and the UN Global Compact (Social Accountability International, 2019):

1. Child labour.
2. Forced labour.
3. Safety and health.
4. Wage compensation.
5. Discrimination.
6. Work discipline.
7. Freedom of association.

8. The right to collective bargaining.
9. Management of the CSR process.

The Standard is applicable to businesses, whether national or international, operating in the Arctic region as it takes into account employees' working conditions.

Nasdaq ESG Reporting Guide

In 2017, Nasdaq initially launched the ESG Reporting Guide, where E represents environmental metrics, S social metrics, and G governance metrics, for private and public companies operating in the Nordic and Baltic countries. Two years later, an improved version was issued as a voluntary global guide, with the aim of operating efficient, fair, and transparent markets and influence effective management of sustainability issues. These will allow Nasdaq, and thus investors, and listed companies to better understand their performance, risks, and resource inefficiency. Serving as a backbone to the ESG Reporting Guide are the EU Directive 2014/95/EU on non-financial reporting (NFRD), the emphasis of the Task Force on Climate-related Financial Disclosure (TCFD), the SDGs, the GRI Standards, and others (Nasdaq, Inc., 2019). Other major sustainability frameworks worth mentioning in the context of the Nasdaq ESG Reporting Guide are the Carbon Disclosure Project (CDP), the Greenhouse Gas Protocol, Sustainability Accounting Standards Board (SASB), the International Integrated Reporting Council (IIRC), and the UN Global Compact (Doni and Johannsdottir, 2019). Table 9.4 lists what ESG issues should be addressed by listed companies. Although initially aimed at Nordic companies, this Guide does not address Arctic-specific issues such as those related to indigenous peoples, although the governance metric, G6 Ethics and Anti-Corruption, may be of much relevance given the potential imbalance in the bargaining power of large multinational corporations and small local communities.

International Labour Organization conventions

The International Labour Organization (ILO) constituents (governments, employers, and workers) prepare and specify international labour standards, fundamental principles, and workers' rights through legally binding international treaties (conventions) that member states can ratify, or through Recommendations that are non-binding guidelines. These are then adopted at the annual ILO conference. The ILO Governing Body has recognized eight fundamental Conventions (International Labour Organization, n.d.a):

1. Freedom of Association and Protection of the Right to Organise Convention, 1948 (No. 87)
2. Right to Organise and Collective Bargaining Convention, 1949 (No. 98)

Table 9.4 ESG reporting guide

Environmental (E) metrics		*Social (S) metrics*		*Governance (G) metrics*	
E1	GHG Emissions	S1	CEO Pay Ratio	G1	Board Diversity
E2	Emissions Intensity	S2	Gender Pay Ratio	G2	Board Independence
E3	Energy Usage	S3	Employee Turnover	G3	Incentivized Pay
E4	Energy Intensity	S4	Gender Diversity	G4	Collective Bargaining
E5	Energy Mix	S5	Temporary Worker Ratio	G5	Supplier Code of Conduct
E6	Water Usage	S6	Non-Discrimination	G6	Ethics & Anti-Corruption
E7	Environmental Operations	S7	Injury Rate	G7	Data Privacy
E8	Climate Oversight / Board	S8	Global Health & Safety	G8	ESG Reporting
E9	Climate Oversight / Management	S9	Child & Forced Labor	G9	Disclosure Practices
E10	Climate Risk Mitigation	S10	Human Rights	G10	External Assurance

Source: Nasdaq (2019, p. 13)

3. Forced Labour Convention, 1930 (No. 29) (and its 2014 Protocol)
4. Abolition of Forced Labour Convention, 1957 (No. 105)
5. Minimum Age Convention, 1973 (No. 138)
6. Worst Forms of Child Labour Convention, 1999 (No. 182)
7. Equal Remuneration Convention, 1951 (No. 100)
8. Discrimination (Employment and Occupation) Convention, 1958 (No. 111)

Although important in the global context, and therefore in the Arctic as well, there are other ILO Conventions that are even more important in the Arctic context: the Indigenous and Tribal Peoples Convention, 1989 (No. 169), and older Indigenous and Tribal Populations Convention, 1957 (No. 107). To date, these Conventions are the only international treaties that exclusively deal with the fundamental rights of indigenous and tribal peoples, establishing "general policies regarding indigenous and tribal peoples' customs and traditions, land rights, the use of natural resources found on traditional lands, employment, vocational training, handicrafts and rural industries, social security and health, education, and cross-border contacts and communication" (International Labour Organization, n.d.b).

The UN Declaration on the Rights of Indigenous Peoples

The United Nations Declaration on the Rights of Indigenous Peoples (UNDRIP) is a non-legally binding resolution (A/61/295) adopted in 2007 by the United Nations General Assembly. The rights of Indigenous peoples are claimed, for instance, in Article 1 stating that "Indigenous peoples have the right to the full enjoyment, as a collective or as individuals, of all 8 human rights and fundamental freedoms as recognized in the Charter of the United Nations, the Universal Declaration of Human Rights and international human rights law" (pp. 7–8). Article 3 recognizes the right to self-determination, including political status, and how Indigenous peoples pursue their economic, social, and cultural development. Article 4 confirms the right to self-government or autonomy in internal and local affairs. Article 5 safeguards the rights to retain and reinforce Indigenous people's own political, legal, economic, social, and cultural institutions. Furthermore, Article 26 claims Indigenous people's legal rights to their traditionally owned, occupied, used or acquired lands, territories, and resources. The rights recognized in UNDRIP constitute according to Article *43* "the minimum standards for the survival, dignity and well-being of the indigenous peoples of the world" (Indigenous foundations, n.d.; United Nations, 2007).

The UN Principles on Businesses and Human Rights

The UN Principles on Businesses and Human Rights (UNGPs) is another instrument of importance as it focuses on businesses and human rights, and thus is of relevance for transnational corporations and other business enterprises operating in the Arctic. It consists of 31 Guiding Principles, including a "Protect, Respect and Remedy" framework on human rights (United Nations, 2011). The Principles recognize the following (United Nations, 2011, p. 1):

a) States' existing obligations to respect, protect and fulfil human rights and fundamental freedoms.
b) The role of business enterprises as specialized organs of society performing specialized functions, required to comply with all applicable laws and to respect human rights.
c) The need for rights and obligations to be matched to appropriate and effective remedies when breached.

For businesses in particular they should respect human rights (Article 11); reference is made to internationally recognized human rights, including the International Bill of Human Rights and the ILO's Fundamental Principles and Rights at Work (Article 12), avoid causing adverse human rights impacts, and business enterprises should prevent or mitigate adverse such impacts (Article 13). The Principles apply to all enterprises regardless of size, sector, operational context, ownership, and structure (Article 14), and enterprises should implement policies and processes that are appropriate to their size and circumstances (Article 15), with the commitment of the most senior-level executives (Article 16). Furthermore, a due diligence investigation should be carried out so that adverse human rights impacts can be identified, prevented, mitigated, and accounted for (Article 17) (United Nations, 2011).

The International Organization for Standardization Standards

The International Organization for Standardization (ISO) offers a range of standards that are of relevance for Arctic conditions. To start with, ISO 26000:2010 (ISO SR) offers a guidance on social responsibility; businesses should conduct their business in an ethical and transparent manner. Implementation of ISO 26000 should contribute to achievement of the UN Sustainable Development Goals, and can help organizations that wish to implement the OECD Guidelines for Multinational Enterprises (ISO, n.d.a). A schematic overview of the standard includes scope, terms and conditions, understanding social responsibility, principles of social responsibility, fundamental practices of social responsibility, the core subjects, and integration of social responsibility throughout an organization, with the purpose of maximizing organizational contribution to sustainable development. The principles of social responsibility are accountability, transparency, ethical behaviour, respect for stakeholder interests, rule of law, international norms of behaviour, and human rights. The fundamental practices of social responsibility are to recognzie social responsibility and stakeholder identification and engagement. The core social reponsibility subjects identified in the standard are human rights, labour practices, the natural environment, fair operating practices, consumer issues, and community involvement and development (ISO, n.d.b), many of critical importance in the Arctic.

Given the fragile Arctic environment, the ISO 14000 family of standards related to environmental management and responsibility is also relevant for businesses operating in the Arctic. Of these the ISO 14001:2015 will be briefly explained, as it sets out the criteria for an environmental management system. The standard is aimed at helping organizations improve their environmental performance. This is done by more efficient use of resources and through waste reduction, which should result in competitive advantage and enhanced stakeholder trust (ISO, 2015). Just as with any other ISO standard the aim is continuous improvement in registered performance.

ISO 45001:2018 is a management system standard focusing on occupational health and safety (previously the BS OHSAS 18001: 2007 Standard), aimed at minimizing the burden of work-related injuries and diseases, thus providing safe and healthy working conditions (ISO, n.d.c). This standard is applicable in the Arctic in view of often harsh working conditions, which can be addressed with effective preventive and protective measures (ISO, n.d.c).

Other relevant standards, or family of standards are ISO 31000:2018 on risk management, and Arctic industry-specific standards on petroleum and natural gas industries.

Carbon Disclosure Project

The Carbon Disclosure Project (CDP) operates a global environmental disclosure system, through which they help companies, cities, states, and regions to measure and manage their risks and opportunities on climate change, water security, and deforestation (CDP, n.d.a). This is of relevance in the Arctic region, such as in terms of business practices, since the region is "extremely vulnerable to current and projected climate change" and has the "greatest potential to affect global climate and thus human populations and biodiversity" (Anisimov et al., 2007, p. 655). On behalf of over 525 institutional investors, with a combined US$96 trillion in assets, CDP request information from the world's largest companies on climate risks and low-carbon opportunities (CDP, n.d.b). One of CDP's areas of research was to identify the 100 major fossil fuel producers and their cumulative emissions (1988–2015). According to the study, they have been collectively responsible for 71% of industrial greenhouse gas emissions since 1988. These companies include some major producers operating in the Arctic region (Griffin, 2017).

Principles for Responsible Investment

The United Nations-supported Principles for Responsible Investment (PRI) offers a blueprint for institutional investors who strive to better understand the implications of environmental, social, and governance (ESG) factors for investment and ownership decisions. Partners to the Principles include the United Nations Environment Programme Finance Initiative (UNEP FI) and the United Nations Global Compact. The PRI's mission is to support long-term value creation benefitting both the environment and society, where the impact areas of the Principles include responsible investors, sustainable markets, and a prosperous world for all. Incorporating ESG factors into investors' decisions and actions should lead to better risk management and enhanced return on investments. PRI signatories commit to the following Principles, thus fostering accountability, integrity, and good governance (PRI Association, n.d.):

1. We will incorporate ESG issues into investment analysis and decision-making processes.
2. We will be active owners and incorporate ESG issues into our ownership policies and practices.
3. We will seek appropriate disclosure on ESG issues by the entities in which we invest.
4. We will promote acceptance and implementation of the Principles within the investment industry.
5. We will work together to enhance our effectiveness in implementing the Principles.
6. We will each report on our activities and progress towards implementing the Principles.

Although, the Principles offer important guidance for institutional investors, they are too broad to offer any Arctic-specific investment guidelines; thus, it is important to mention briefly others, including the Arctic Investment Protocol.

The Arctic Investment Protocol

The Arctic Investment Protocol from 2016 was a pioneering effort of Guggenheim, a global investment and advisory financial services firm. The aim was to advance responsible and sustainable business practices in the Arctic. The Protocol was then transferred by the World Economic Forum to the Arctic Economic Council (ACE), an independent organization that facilitates Arctic business activities and responsible economic development (Guggenheim Investments, 2020).

The Arctic Investment Protocol offers six Principles for promoting equitable and sustainable economic growth in the Arctic that should support the well-being and resilience of local communities, based on fairness, inclusiveness, and environmentally sound practices. The Principles Arctic investors should follow are (WEF, n.d.):

1. Build resilient societies through economic development.
2. Respect and include local communities and indigenous peoples.
3. Pursue measures to protect the environment of the Arctic.
4. Practice responsible and transparent business methods.
5. Consult and integrate science and traditional ecological knowledge.
6. Strengthen pan-Arctic collaboration and sharing of best practices.

In developing the Arctic Investment Protocol, some major international guidelines and institutions were consulted, some of which are discussed in this section. These are the Aspen Principles for Long-Term Value Creation, the Aspen Principles of Arctic Governance, the Equator Principles, the Extractive Industries Transparency Initiative (EITI), the International Council on Mining & Metals Indigenous Peoples and Mining Position Statement, the International Maritime Organization (IMO), the International Finance Corporation (IFC) Performance Standards on Environmental and Social Sustainability, the National Science Foundation (U.S.) Principles for the Conduct of Research in the Arctic, the Organisation for Economic Co-operation and Development Guidelines for Multinational Enterprises, the UN Guiding Principles on Business and Human Rights, the UN Principles for Responsible Investment, and the World Bank Group Environmental, Health and Safety Guidelines (WEF, n.d.).

Further resources

Other tools, principles, guidelines, and standards may be of relevance for businesses operating in the Arctic. The European Commission offers an employment and recruitment agencies sector guide on implementing the UN guiding principles on business and human rights, which in particular mentions potentially vulnerable or marginalized individuals and groups, including Indigenous Peoples, and thus refers to the UN Declaration on the Rights of Indigenous Peoples (European Commission, 2014a). The European Commission has also developed sector-specific guidance for the oil and gas sector, titled the *Oil and Gas Sector Guide on Implementing the UN Guiding Principles on Business and Human Rights*. The Guiding Principles also recognizes the rights of individuals or groups in a position of heightened vulnerability or marginalization, including Indigenous Peoples. The analytical framework for assessing potential impacts of company activities on stakeholder groups recognizes, in relation to Indigenous People's rights: land acquisition and resettlement and free, prior and informed consent (FPIC); drilling and seismic testing; and external communication and consultation which may be complicated due to language barriers. The Guiding Principles furthermore offers a human rights risk map for prioritizing action, where risks are evaluated based on likelihood and severity (European Commission, 2013). Yet another Guide developed by the European Commission is the information and communications technologies (ICT) sector guide on implementing the UN Guiding Principles on Business and Human Rights, which briefly addresses Indigenous People‘s rights in relation to component and network equipment manufacturing (European Commission, 2014b). The human rights of Indigenous Peoples are furthermore identified in relation to the operation of small and medium-sized enterprises (SMEs), specifically the right to self-determination and land rights (European Commission, 2012).

The European Commission offers a similar management and audit scheme to the ISO. This is the EU Eco-Management and Audit Scheme (EMAS) developed for companies and other organizations to evaluate, report, and improve their environmental performance. ISO 14001 can be used as a step up to EMAS as EMAS has more stringent requirements than ISO 14001. The regulation was first introduced in 1993 as EMAS Regulation 1836/93, and was later revised in 2001 as Regulation (EC) No 761/2001 ('EMAS II') and in 2009 as Regulation (EC) No 1221/2009 ('EMAS III'). The EMAS Global mechanism enables companies with sites outside Europe to employ EMAS (European Commission, n.d.b). The difference between EMAS and ISO14001 lies for instance in an initial environmental review of all direct and indirect environmental aspects, which is required in EMAS but not according to ISO 14001. According to EMAS, there is a focus on continual improvement of environmental performance, but according to ISO 14001 there is a focus on continual improvement of the management system. EMAS also requires proof of legal compliance and external reporting and open dialogue with stakeholders, while ISO 14001 does not requires compliance with audit or external communication (EMAS Info, 2014).

The International Council on Mining and Metals has issued an Indigenous Peoples and mining good practice guide. This guide is first and foremost used to influence good practices where mining-related activities take place on or near traditional indigenous land and territory. The focus of the guide is on principles of good engagement and indigenous participation, including free, prior, and informed consent, how to manage impact, making agreements, and how to deal with concerns and grievances in indigenous communities. The guide also offers various practical tools for mining and metal companies to apply (International Council on Mining and Metals, 2015).

Case study: The Danish CSR Compass

The CSR Compass is a Danish website developed in collaboration between the Danish Ministry of Business and Growth and the Confederation of Danish Industry; see www.csrcompass.com/other-certification-schemes. The Danish Institute for Human Rights (IMR) is, furthermore, a strategic partner for the CSR Compass. In addition, in the latest revision a dialogue with Danish Ethical Trading Initiative (DIEH) regarding the tool content and form was carried out, in addition to an input from other Nordic stakeholders, namely (CSR Compass, n.d.a):

- From Iceland: The Confederation of Icelandic Employers (SA)
- From Finland: Ministry of Employment and the Economy
- From Norway: Ministry of Foreign Affairs, the Agency for Public Management and Government, and the Confederation of Norwegian Enterprise (NHO)
- From Sweden: The Swedish Environmental Management Council and the International Chamber of Commerce

The purpose of the CSR Compass is to guide companies in the right direction by offering information about, and links to, various well-known national and international tools useful to companies in their CSR work, including reporting. According to the CSR Compass, companies should employ due diligence in their CSR activities, both in terms of their own business affairs and activities and also in relations with stakeholders, including employees, customers, suppliers, and investors to name a few (CSR Compass, n.d.b). The purpose of the due diligence is to identify areas where CSR principles might be violated (by doing so companies can prevent and mitigate potential adverse impacts) and to report on how such impacts were handled. The CSR Compass also emphasizes responsible supply chain management to minimize risks of negative public relations while securing buy-in from investors and creditors and/or creating new market opportunities. This is done by ensuring in a collaborative manner that suppliers have CSR strategies in place and that their performance is being monitored for improvements. The focus should be on high-risk suppliers, such as in particular industries suppliers operating in countries with weak environmental and social legislation, or in countries where internationally recognized principles are not fully respected, such as in the areas of human rights and labor rights, with regard to the natural environment or when anti-corruption measures are not taken (CSR Compass, n.d.c).

The CSR Compass recognizes principles, guidelines, and standards as an important starting point for companies that want to instigate and develop further their CSR activities. Principles, systems, and relevant indicators for CSR strategy, implementation, and reporting may come from Danish and international organizations, some of which may lead to certification (CSR Compass, n.d.d). The CSR Compass also offers links to guidelines and some self-assessment tools for compliance such as the Global Compact Self-Assessment Tool (n.d.) and the Human Rights Compliance Assessment (HRCA) Quick Check (The Danish Institute for Human Rights, 2016). Furthermore, companies are urged to establish an effective complaint mechanism, either themselves or in collaboration with other parties, to deal with conflict cases (CSR Compass, n.d.e).

Questions for discussion

1. What tools, principles, conventions, guidelines, and standards does the Danish CSR Compass recognize?
2. How relevant are these tools, principles, conventions, guidelines, and standards for Arctic operation and reporting?
3. What critical tools, principles, conventions, guidelines, and standards of relevance for reporting are missing, if any?
4. What benefits, or limitations, do you recognize of employing the Danish CSR Compass?

Chapter summary

This chapter has analysed the tools, conventions, principles, guidelines, and standards that are recognized as fundamental for auditing and reporting on Arctic operations. These are the UN Global Compact, the Global Reporting Initiative, the OECD Guidelines for Multinational Enterprises, the Sustainability Accounting Standards Board, the International Integrated Reporting Council, AccountAbility, Social Accountability 8000, Nasdaq ESG Reporting Guide, the International Labour Organization (ILO) conventions, the UN Declaration on the Rights of Indigenous Peoples, the UN Principles on Businesses and Human Rights, the International Organization for Standardization Standards, the Carbon Disclosure Project, UN Principles for Responsible Investment, and the Arctic Investment Protocol. Some other tools and guidelines have been mentioned, mainly guidelines devolved by the European Commission, the European Eco-Management and Audit Scheme, and a good practice guide from the International Council on Mining and Metals. The case study in this chapter is the Danish CSR Compass, a tool developed to guide companies in the right direction by selecting national and international tools useful to companies in their CSR work, including auditing and reporting.

This analysis highlights the complexity of the Arctic operational condition, and therefore auditing and reporting. Multiple tools, conventions, principles, guidelines, and standards need to be considered in order to ensure Indigenous People's rights. This makes comparison between companies' reports very difficult, if at all possible. Yet another issue that needs to be addressed are key performance indicators (KPIs), which are specific performance measurements used to evaluate how successfully companies are achieving their business objectives and operational goals. They can both be qualitative and quantitative, as well as leading (predictive measurement) and lagging (output measurements). In general, the tools, conventions, principles, guidelines, and standards give limited or no guidance on Arctic-related KPIs, meaning it is more or less impracticable to recognize whether organizational practices in the Arctic are improving or not. Some of the tools and guidelines in the chapter are, furthermore, generic but not Arctic specific, including the UN Global Compact, the Global Reporting Initiative and others. The benefit of using these tools is nevertheless of importance but they cannot be used without considering other relevant tools, conventions, principles, guidelines, and standards.

References

AccountAbility. (2018a). AA1000 Accountability Principles 2018. Accessed on 25 February 2020 from www.accountability.org/standards/

AccountAbility. (2018b). AA1000 Assurance Standard (AA1000AS, 2008) with 2018 addendum. Accessed on 25 February 2020 from www.accountability.org/standards/

AccountAbility. (2018c). AA1000 Stakeholder Engagement Standard 2015. Accessed on 25 February 2020 from www.accountability.org/standards/

AccountAbility. (2020). Standards. Accessed on 25 February 2020 from www.accountability.org/standards/

Anisimov, O.A., Vaughan, D.G., Callaghan, T.V., Furgal, C., Marchant, H., Prowse, T.D., Vilhjálmsson H., and Walsh, J.E. (2007). Polar regions (Arctic and Antarctic). In M.L. Parry, O.F. Canziani, J.P. Palutikof, P.J. van der Linden and C.E. Hanson (Eds.), *Climate Change 2007: Impacts, Adaptation and Vulnerability. Contribution of Working Group II to the Fourth Assessment Report of the Intergovernmental Panel on Climate Change* (pp. 653–685). Cambridge, UK: Cambridge University Press.

Arctic Council. (2013). Kiruna Declaration. Accessed on 14 February 2020 from https://oaarchive.arctic-council.org/handle/11374/93

Arctic Economic Council. (n.d.). Arctic Investment Protocol. Guidelines for Responsible Investment in the Arctic. Accessed on 25 February 2020 from www.guggenheiminvestments.com/cmspages/getfile.aspx?guid=3be60aa9-5229-4049-9bfe-22dfab0be9ef

Arruda, G.M. (2019). *Sustainable Energy Education in the Arctic: The Role of Higher Education* (1st ed.). Abingdon, UK: Routledge.

CDP. (n.d.a). What we do. Accessed on 25 February 2020 from www.cdp.net/en/info/about-us/what-we-do

CSR Compass. (n.d.a). The parties behind the initiative. Accessed on 14 February 2020 from www.csrcompass.com/parties-behind-initiative

CSR Compass. (n.d.b). Keeping your house in order. Accessed on 14 February 2020 from www.csrcompass.com/keeping-your-own-house-order

CSR Compass. (n.d.c). Responsible supply chain management. Accessed on 14 February 2020 from www.csrcompass.com/responsible-supply-chain-management

CSR Compass. (n.d.d). Guidelines. Accessed on 14 February 2020 from www.csrcompass.com/guidelines

CSR Compass. (n.d.e). Complaint mechanisms. Accessed on 14 February 2020 from www.csrcompass.com/complaint-mechanisms

Doni, F. and Johannsdottir, L. (2019) Environmental social governance (ESG) ratings. In W. Leal Filho, A. Azul, L. Brandli, P. Özuyar, and T. Wall (Eds.) *Encyclopedia of the UN Sustainable Development Goals* (pp. 1–15). Cham, Switzerland: Springer.

EMAS Info. (2014). *From ISO 14001 to EMAS: Mind the Gap*. Berlin: Author. Accessed on 20 December 2020 from www.emas.de/fileadmin/user_upload/4-pub/UGA_Infosheet_From-ISO-14001-to-EMAS.pdf

European Commission. (2012). *My Business and Human Rights: A Guide to Human Rights for Small and Medium-sized Enterprises*. Brussels: Author. Accessed on 25 February 2020 from www.business-humanrights.org/sites/default/files/documents/SME-BHR-guide-EU.pdf

European Commission. (2013). *Oil and Gas Sector Guide on Implementing the UN Guiding Principles on Business and Human Rights*. Brussels: Author. www.ihrb.org/pdf/eu-sector-guidance/EC-Guides/O&G/EC-Guide_O&G.pdf

European Commission. (2014a). *Employment & Recruitment Agencies Sector Guide on Implementing the UN Guiding Principles on Business and Human Rights*. NB-04-13-164-EN-N. Brussels: Author.

European Commission. (2014b). *ICT Sector Guide on Implementing the UN Guiding Principles on Business and Human Rights*. NB-04-13-165-EN-N. Brussels: Author.

European Commission. (2017). *Commission Guidelines on Non-financial Reporting (Methodology for Reporting Non-financial Information)*. Accessed on 14 February 2020 from https://eur-lex.europa.eu/legal-content/EN/TXT/?uri=CELEX%3A52017XC0705%2801%29

European Commission. (2019). *Guidelines on Non-financial reporting: Supplement on Reporting Climate-related Information*. Accessed on 14 February 2020 from https://ec.europa.eu/finance/docs/policy/190618-climate-related-information-reporting-guidelines_en.pdf

European Commission. (n.d.a). Non-financial reporting. Accessed on 14 February 2020 from https://ec.europa.eu/info/business-economy-euro/company-reporting-and-auditing/company-reporting/non-financial-reporting_en

European Commission. (n.d.b). Policy – EMAS regulation. Accessed on 28 February 2020 from https://globalcompactselfassessment.org/

Global Compact Self Assessment Tool. (n.d.). Global Compact Self Assessment Tool. Accessed on 14 February 2020 from https://globalcompactselfassessment.org/

Global Reporting Initiative (GRI). (n.d.a). About GRI. Accessed on 14 February 2020 from www.globalreporting.org/Information/about-gri/Pages/default.aspx

Global Reporting Initiative (GRI). (n.d.b). GRI's history. Accessed on 14 February 2020 from www.globalreporting.org/about-gri/mission-history/#:~:text=GRI%20was%20founded%20in%20Boston,of%20the%20UN%20Environment%20Programme)

Global Reporting Initiative (GRI). (n.d.c). GRI standards download centre. Accessed on 14 February 2020 from www.globalreporting.org/standards/gri-standards-download-center/

Global Reporting Initiative (GRI). (2018a). GRI 411 Rights of indigenous peoples 2016. Amsterdam: Author.

Global Reporting Initiative (GRI). (2018b). *The Materiality Principle: A Deep Dive*. Accessed on 25 February 2020 from https://vdocuments.mx/the-materiality-principle-a-deep-dive-2018-12-04-the-way-we-design-produce-and.html

Griffin, P. (2017). *The Carbon Majors Database: CDP Carbon Majors Report 2017*. London: CDP.

Guggenheim Investments. (2020). The Arctic Investment Protocol. Accessed on 25 February 2020 from www.guggenheiminvestments.com/institutional/firm/sustainable-investing-esg/the-arctic-investment-protocol

Indigenous foundations. (n.d.). UN Declaration on the Rights of Indigenous Peoples. Accessed on 28 February 2020 from https://indigenousfoundations.arts.ubc.ca/un_declaration_on_the_rights_of_indigenous_peoples/

International Council on Mining and Metals. (2015). *Good Practice Guide. Indigenous Peoples and Mining* (2nd ed). London: Author. Accessed on 28 February 2020 from www.icmm.com/website/publications/pdfs/social-performance/2015/guidance_indigenous-peoples-mining.pdf

International Integrated Reporting Council (IIRC). (n.d.). Integrated Reporting. Accessed on 25 February 2020 from https://integratedreporting.org/ir-business network/

International Labour Organization (ILO). (n.d.a). Conventions and recommendations. Accessed on 28 February 2020 from www.ilo.org/global/standards/introduction-to-international-labour-standards/conventions-and-recommendations/lang--en/index.htm

International Labour Organization (ILO). (n.d.b). International labour standards on Indigenous and tribal peoples Accessed on 28 February 2020 from www.ilo.org/global/standards/subjects-covered-by-international-labour-standards/indigenous-and-tribal-peoples/lang--en/index.htm

ISO. (n.d.a). ISO 26000 Social Responsibility. Accessed on 25 February 2020 from www.iso.org/iso-26000-social-responsibility.html

ISO. (n.d.b). Schematic overview of ISO 26000. Accessed on 25 February 2020 from www.iso.org/files/live/sites/isoorg/files/store/en/PUB100260.pdf

ISO. (n.d.c). ISO 45001 Occupational health and safety. Accessed on 25 February 2020 from www.iso.org/iso-45001-occupational-health-and-safety.html

ISO. (2015). ISO 14001. Accessed on 25 February 2020 from www.iso.org/files/live/sites/isoorg/files/store/en/PUB100371.pdf

KPMG. (n.d.). The rise of responsible investment: Four recommendations institutional investors and companies should consider when building ESG policies. Accessed on 14 February 2020 from https://home.kpmg/xx/en/home/insights/2019/03/the-rise-of-responsible-investment-fs.html

Nasdaq, Inc. (2019). *ESG Reporting Guide 2.0: A Support Resource for Companies*. New York: Author. Accessed on 16 February 2020 from www.nasdaq.com/docs/2019/11/26/2019-ESG-Reporting-Guide.pdf

OECD Watch. (n.d.). The OECD Guidelines. Accessed on 28 February 2020 from www.oecdwatch.org/oecd-ncps/the-oecd-guidelines-for-mnes/

OECD Publishing. (2017). *OECD Due Diligence Guidance for Meaningful Stakeholder Engagement in the Extractive Sector*. Paris: Author.

Organisation for Economic Co-operation and Development (OECD). (2011). *OECD Guidelines for Multinational Enterprises*. Accessed on 28 February 2020 from www.oecd.org/daf/inv/mne/48004323.pdf

Organisation for Economic Co-operation and Development (OECD). (2017). *Responsible Business Conduct for Institutional Investors: Key Considerations for Due Diligence under the OECD Guidelines for Multinational Enterprises*. Accessed on 28 February 2020 from https://mneguidelines.oecd.org/RBC-for-Institutional-Investors.pdf

Organisation for Economic Co-operation and Development (OECD). (2018). *OECD Due Diligence Guidance for Meaningful Stakeholder Engagement in the Extractive Sector*. Accessed on 28 February 2020 from https://mneguidelines.oecd.org/stakeholder-engagement-extractive-industries.htm

PRI Association. (n.d.). *Principles for Responsible Investment: An Investor Initiative in Partnership with UNEP Finance Initiative and the UN Global Compact*. Accessed on 25 February 2020 from www.unpri.org/download?ac=6303

Social Accountability International (SAI). (2019). SA8000® Standard. Accessed on 25 February 2020 from www.sa-intl.org/index.cfm?fuseaction=Page.ViewPage&PageID=1689

Sustainability Accounting Standards Board (SASB). (n.d.). Standards overview. Accessed on 25 February 2020 from www.sasb.org/standards-overview/

Sustainability Accounting Standards Board (SASB). (2018a). Why SASB? Accessed on 25 February 2020 from www.sasb.org/why-sasb/

Sustainability Accounting Standards Board (SASB). (2018b). Materiality Map®. Accessed on 25 February 2020 from https://materiality.sasb.org/

Sustainable Development Working Group (SDWC). (2013). Corporate social responsibility in the Arctic. Accessed on 14 February 2020 from www.sdwg.org/corporate-social-responsibility-in-the-arctic/

The Danish Institute for Human Rights. (2016). Human rights compliance assessment quick check. Accessed on 14 February 2020 from www.humanrights.dk/publications/human-rights-compliance-assessment-quick-check

The International Integrated Reporting Council. (2013). The International <IR> Framework. Accessed on 25 February 2020 from https://integratedreporting.org/wp-content/uploads/2015/03/13-12-08-THE-INTERNATIONAL-IR-FRAMEWORK-2-1.pdf

UN Global Compact. (n.d.). The power of principles. Accessed on 14 February 2020 from www.unglobalcompact.org/what-is-gc/mission/principles

UN Global Compact. (2014). *Guide to Corporate Sustainability: Shaping a Sustainable Future*. New York: Author. Accessed on 14 February 2020 from www.unglobalcompact.org/docs/publications/UN_Global_Compact_Guide_to_Corporate_Sustainability.pdf

UN Global Compact & Deloitte. (2010). *UN Global Compact Management Model: Framework for Implementation*. New York: Ibid. Accessed on 14 February 2020 from www.unglobalcompact.org/docs/news_events/9.1_news_archives/2010_06_17/UN_Global_Compact_Management_Model.pdf

United Nations. (2007). United Nations Declaration on the Rights of Indigenous Peoples. Accessed on 28 February 2020 from www.un.org/development/desa/indigenouspeoples/wp-content/uploads/sites/19/2018/11/UNDRIP_E_web.pdf

United Nations. (2011). *Guiding Principles on Business and Human Rights*. Accessed on 28 February 2020 from www.ohchr.org/Documents/Publications/GuidingPrinciplesBusinessHR_EN.pdf

World Economic Forum (WEF). (n.d.). *Arctic Investment Protocol: Guidelines for Responsible Investment in the Arctic*. Accessed on 25 February 2020 from www3.weforum.org/docs/WEF_Arctic_Investment_Protocol.pdf

10 CSR education for the 21st century

Learning objectives

- Understand the relationship between CSR, sustainability, and education.
- Understand how to manage change and educate for social change.
- Understand how to educate for change in business and CSR.
- Understand How to connect this adaptive education to the business context and ethical issues.
- Understand to promote curriculum and pedagogical changes to promote responsible multicultural interactions and form responsible citizens.

  To involve business and university in this task.
  To involve students with the community needs.

- Understand how to form global citizens capable of tackling the challenges of the 21st century.

Introduction

The speed and scale of global changes makes it hard to predict what Arctic and non-Arctic societies will face in 10 to 50 years' time. Currently, it is not possible to determine completely the future vulnerable spots, or the scale of outcomes generated by the combination of climate change, energy demand, pollution, and human activity. It is also an *incognita* how this set of risks will affect the ecosystem's ability to adapt to complex and dynamic conditions as well as the adaptability of the social environment of Arctic societies to this rapid transition. This level of complexity and unpredictability (Arruda, 2019) indicates the need to reflect on the level of understanding that stakeholders, as players of reality, have and will have about development in a changing world and how they will manage this change effectively.

A sustainable society can be defined as that in which its people have a strong sense of initiative and solidarity, with a solid capacity for mobilization and awareness of rights. They also have a remarkable community organization and a sense of self-management, with a high level of sensitivity to cultural, social, economic, and environmental issues. A sustainable society requires individuals that can think beyond individual interests and consider societal interest. This change in perspective may have a substantial impact on the way sustainability and social responsibilities are perceived, practised, implemented, and embedded in society.

Arctic societies are essentially social. The sense of community and attachment to place are strong characteristics that promote connectivity between the natural and social worlds and empower the young Arctic generation. The social aspects of these societies also connect knowledge to its reality as elderly members of a community are considered the knowledge keepers of Traditional Knowledge that is transmitted by art, language, history, music, medicine, etc. (Battiste and Henderson, 2000). For an indigenous student, these are the primary sources of Traditional Knowledge and an innate mechanism of learning (Arruda, 2019). Indigenous students debate and analyse contexts under a cultural, environmental, social, and therefore holistic perspective. A more sustainable society will need to be naturally inclusive and will require new ways of thinking about communication, knowledge, learning, resilience, and knowledge application by reorienting the narratives to a more intercultural and inclusive dialogue as a global and systemic process aligned with the social, cultural, environmental, and economic pillars of sustainability (Arruda, 2019, p. 65).

Types of discussion and issues explored in Chapter 10

This chapter provides a short overview of the complexity of the dynamics of the Arctic as a region in development and in constant modernization, which impacts its environmental and social realities. This explains the importance of educating for change as well as building bridges between business, social sciences, and education. It also suggests the necessity of convergence and consensus in multicultural interactions (Arruda, 2019). This chapter aims to examine the relationship between corporate social responsibility (CSR), sustainability, and education; and how to manage change and educate for social change in a holistic way but also with a focus on the fields of business and CSR. This is achieved by connecting this adaptive education to the business context and practical ethical issues, taking into consideration the differences between CSR in developing and developed countries. The chapter also considers mechanisms to promote curriculum and pedagogical changes to promote responsible social and multicultural interactions and form responsible global citizens (Peters et al., 2008; Arruda, 2019) capable of tackling the challenges of the 21st century.

In order to tackle these aspects of a modern CSR education, it is also important to understand which key processes of active, applied, and participatory learning (Arruda, 2019) can stimulate critical thinking and views of more sustainable futures based on social responsibility. This is achieved through innovative strategies of reorienting not only education but also communication by stimulating inclusive debates in support of sustainable development. A continuous flow of innovative strategies is necessary in the field of education and will contribute to building capacities into society to understand, in more depth, the transition process; how to approach and manage change through consistent, meaningful, and applied critical reflection on values; and new perspectives to improve the current models of development (Arruda, 2019), and therefore society's quality of life at local, regional, and international levels.

Background: Relationship between CSR, sustainability, and education

The Arctic has contradictions and complexities, and needs to be understood as a complex system of socio-environmental interactions. It is a specialized system formed by physical and human interconnected dimensions; sustainability depends on the balance of these systemic components. A sustainable Arctic depends on the capacity of enhancing the complexity of governance and education. To move to this point, however, citizens and corporations need to understand in theory and practice the concept of social responsibility, recognizing the challenges in perceiving and practising the concept. At the same time, sustainability is also a controversial concept that proved to be challenging to understand in-depth by different societies.

Social responsibility is a commitment to the well-being of others, a naturally social inclusive idea because it is not possible to be collectively responsible if citizens are not responsible as individuals. It is a concept that exists in different dimensions that are intrinsically connected. If we think at this level, it is possible to apply the concept to a corporation, a society, a university, or any organization. The EU CSR focal points align with those in the ISO 26000 guidelines where social responsibility is defined as a "responsibility of an organization for the impacts of its decisions and activities on society and the environment", but should not each citizen have the same concerns? The problem is how to inculcate social responsibility at the level of citizens and cultures. It certainly presents huge challenges for communication and education, because we will have to rethink education and train educators.

Over history, we have got used to the concept of CSR and many theories associated with this concept, but we are now at the point where it is important to practise social responsibility at the level of the individual to reach the collective. For this reason, it is essential to rethink the core of the concept and interrelations between elements of the concept, in order to project this fundamental understanding at the level of citizens to collectives. The same phenomenon applies to the concept of sustainable development. The jargon is well known but in practice means different things to different people (Arruda, 2019). It also needs to be embedded in education to have a meaningful effect on reality.

Sustainable development and education have an intrinsic, interdependent relationship, largely because no sustainable society can operate without high-quality education and human development is intrinsically linked to education standards (AHDR, 2014; Popple, 2015, p. 67; Arruda, 2019). Education needs to be seen beyond attaining degrees but as a way of inculcating and providing live experiences on how students interact and engage with their constructed reality both from an individual and collective perspective, considering individual and societal interests. This is probably the necessary alignment to achieve concrete results within this relationship between CSR, sustainable development, and education.

It is certainly a process that involves values emerging from individuals to societies and from societies to form the next generation of individuals. We need to debate and develop this new level of relationship between individual ethics, collective ethics, and business ethics.

Adaptive education for the business context and ethical issues: Global and local perspective

Adaptive strategies are the ways in which individuals, households, and communities change their productive activities and modify local rules and institutions to secure livelihoods (Berkes and Jolly, 2002, p. 18). The flexibility of this more modern and resilient way of thinking and operating, in practice, allow communities to address climatic uncertainty and constantly adapt to change, in real time, by using information systems to disseminate information within their communities. Adaptation depends on efficient communication that improves with a well-structured education and curriculum oriented to sustainable modes of development. The holistic dimension of Arctic issues requires an equally holistic approach to education, enabling information to be disseminated in a variety of formats intelligible to different stakeholders. Communication and education are really powerful adaptation tools in a changing Arctic (Arruda, 2019).

In the present era, information is volatile and credible means of communication and dissemination are vital for consolidating good practices, good governance, and democratic processes for the well-being of citizens, societies and economies. Therefore, clear communication and dissemination of business ethics have a vital role to play in local and global contexts. Not only because there is a climate emergency underway and values, principles, and ethics are in check around the world, but because the divergent discourses of our times make it more difficult to establish a common dialogue about balanced practices in relation to corporate social responsibility and sustainability. In practical terms, the patterns of western corporate sustainability present dissimilarities in developing and emerging nations. Consequently, the perception of the standards for corporate sustainability performance in emerging and developing countries have different characteristics and trends with adverse effects on sectors of society and economy. Empirical studies and observations based on international audiences approached as part of the methodology of the reflection in this chapter showed that these differences also have huge implications at the levels of governance, communication, knowledge application, and decision-making, shown in the practices and performance indices in developed, developing and emerging economies. Several local contexts have been dragged down by corruption, environmental negligence, and social issues. The current systems of business ethics, corporate governance, and global governance do not discuss the causes but only the consequences when events have already happened. Additionally, this limited discussion extends this issue to business environments. When western higher education students come to engage in western higher education courses and try to learn CSR principles or how it works in business ethics, the CSR learning journey seems to face challenges in terms of communication and comprehension due to the unfamiliarity of concepts and practices. Consequently, the challenges for an adaptive education are significant and require in-depth analysis of the opportunities available if we look at different perspectives. Within an international perspective, this phenomenon can be explained by the difficulty in balancing CSR in developed economies and in the rest of the world. The challenges to understand business ethics in the developing world are higher than those in the developed world. The principles, values, and ethics seem to be difficult to perceive and developing economies see CSR in a different way because these values, ethics, and principles are not embedded in the mindsets of citizens and societies. Additionally, CSR is commonly approached from a philanthropic viewpoint and publication systems privilege the dominant perspective of CSR. In communicating CSR, values, principles, and ethics are not part of the mainstream agenda in the developing world. These components (CSR principles, values, and ethics) are not embedded in the reality of several countries and their respective societies. The role of communication and dissemination of knowledge and information about the local dynamics of developing and emerging economies and business practices can raise important points for shared global debate by professionals and students. Moreover, in some of these countries the debate is suppressed by the few or inexistent sources of communication and publications reflecting the low stimulus for extensive debate about local contexts and case studies. Ethics and corporate sustainability are difficult to perceive as part of the culture in some jurisdictions, because it is not part of the local debate and there are very few or, sometimes, no organizations engaged in discussing these issues locally and globally. If scholars from developing and emerging nations are given the same opportunity as western scholars to communicate their ideas, contexts, experiences, and realities in terms of business ethics and corporate sustainability, then a state of balance will be reached. Local goals will become embedded into organizations and these more sustainable practices and patterns will have a better chance to influence and, possibly, transform governmental policies and governance systems.

Within this context, the Arctic is a particular case due to the specificity of the socio-cultural-environmental dynamics. As a region with much current and continuous development, the impact of business and entrepreneurism in the

Arctic in different industries may generate a wave of innovation in technology, international politics, and governance that will redefine the regional identity, or what is called Arctic Citizenship (Arruda, 2014, 2015, 2019). Increased industrial activity in the North further exacerbates the already severe climatic impacts, which include changes in weather patterns, ecosystems, as well as the access, availability, and quality of food and water (Willox et al., 2013), affecting the physical and the socio-cultural landscape of the Arctic (Arruda and Krutkowski, 2017). Climate change and modernization are two intrinsically linked forces altering the current context in which the indigenous populations sustain a livelihood (van Voorst, 2009, p. 236).

Living in and off the richness of complex Arctic ecosystems, indigenous populations have an in-depth understanding of the functioning of ecosystems and the techniques for using and managing them in a sustainable way. Their traditional values, activities, and practices contributed to the development of unique economies and community lifestyles. The indigenous people's attachment to the land is an essential component of their cultural habits which influence their intergenerational knowledge and values transmission, and consequently their identity and representativeness (Arruda and Krutkowski, 2016).

Arctic indigenous communities understand knowledge according to an indigenous system of conceptualization, classification, methodology, and analysis based on herding diaries, snow and temperature measurements (Eira, 2012, p. 129), and the application of physical and strategic knowledge components of risk assessment and adaptation. According to this viewpoint, 'knowledge' and 'reality' should not be understood as separate systems, but as part of the same emerging complex system (Boulton et al., 2015, p. 11, 20) that allows them to assess risk in the field. This is how they build up their intergenerational knowledge systems (Arruda, 2019).

The CSR and sustainability debate in the Arctic requires exploration of challenges and adaptation mechanisms through multiple knowledge systems (Bitz et al., 2016; Arruda, 2018) by involving students and scholars in inclusive and democratic debates based on their experiences and perspectives. Active social participation in the local climate and sustainability debate depends on efficient sustainability communication (Franz-Balsen and Heinrichs, 2007; Gustafson, 2016) and literacy initiatives: energy literacy, climate literacy, and social responsibility literacy with practical application in the field.

It is important to decolonize the discourse, communication, and curriculum to address the contradictions of sustainable development and social responsibility. Modern Arctic education requires knowledge and information to be co-created and disseminated in a variety of formats intelligible to a variety of stakeholders. These guidelines can also be applied to an innovative global education. The Arctic is the arena where these different world systems encounter and demand adaptations to thrive. It is a natural stage for intercultural knowledge, information, and innovation.

Curriculum and pedagogical changes to promote socially responsible interactions

There is little empirical research about CSR in developing countries despite the literature often framing the debate of CSR within a global context. In developing countries around the world, CSR is still seen as a minimal commitment to social responsibility in the face of the economic pillar. In several developing countries, it is not seen as a strategic commitment. In several countries, CSR is at different stages of development and commitment, which explains the different perspectives and levels of adherence to its precepts that determine how it is perceived and understood. The perceptions of students and the public on CSR vary according to the levels of alignment and embeddedness with principles and practices adopted in different developing countries.

Only recently has the emphasis on closer alignment between socio-environmental components and the curriculum started to gain attention. As citizens, the difficulty in understanding and applying the concepts of social responsibility and sustainable development is explained by the lack of adherence of the education curriculum to these concepts, both theoretically and practically. The developmental model of curriculum planning suggests that we should look to a particular view of humanity and human development, including social development and cultural components (Kelly, 2009, p. 88, 99), because cultural difference contributes to the managerialist and organizational goals of development (Radcliffe, 2006, p. 237). It means that the concept and features of an Arctic Citizenship (Arruda, 2019) and a Global Citizenship (UNESCO, 2015, p. 15); Pittman, 2017) should be nurtured in the curriculum according to clear criteria. Curricula should not only focus on key competencies for the world of paid employment, because this perception is deficient (Marsh, 2009, p. 7; Arruda, 2019). Learners are citizens, not only industry professionals (Arruda, 2019).

This curriculum alignment also depends on the recognition of a necessary shift from social responsibility as a minimal commitment (Chandler and Werther, 2013) to this becoming a leading strategic driver of change, participation, and inclusion. CSR evolved in different ways in different countries, but the citizenship sense of responsibility in education (Arruda, 2019) should be based on the SDGs as clear parameters of development.

The Arctic as a developing region presents an enormous advantage in terms of applying, integrating, and implementing CSR because the social-environmental aspects are embedded into Arctic communities' cultural systems and Traditional Knowledge. The way CSR and sustainability education are approached should ideally take into account the context of the community/society, the economic activities, and the university function in terms of content, methods, and pedagogies. CSR also depends on risk assessment and aspects of Arctic business operations and local needs. The framework should be considered as holistic and integrated.

The importance of the social dimension should be part of the day-to-day academic activities and engaging students in community service should become part of the integrated curriculum. Such strategic educational inputs will help students to build a sense of responsibility and citizenship, and consequently a more sustainable future based on an efficient operational environment. Educators and students should be co-creators of a curriculum that should be flexible and subject to timely reviews in order to address the pressing needs of the moment.

Universities should join forces and engage in social responsibility and engaged citizenship education with a clear focus, policy, and strategy to promote socially responsible education. Embedding social responsibility into the curriculum provides a new ethos to education; therefore, the process of curriculum development is of great importance not only to educational institutions but also to the entire society because it stipulates the educational policy and direction of local governance. There is an urgent need for an innovative educational system driven by the desire to provide education relevant to collective needs, distributive of social and economic benefits that leads to enhancement of individuals' living standards and overall societal transformation. The future rests on how the educational sector could produce individuals who can positively impact their communities by embracing international best practices. Innovating curricula will facilitate the inculcation of social responsibility into students' minds when they are active participants of social responsibility and this value is embedded into the educational system. This is one facet of the transformative aim of higher education (Tilbury, 2011, p. 3) and an important component of sustainable development literacy.

How to form global citizens capable of tackling the challenges of the 21st century

The Arctic is a specialized system formed of physical and human interconnected dimensions. Environmental and social systems co-exist in constant interaction. The physical subsystems are formed by glaciers, areas of tundra, mountains, fjords, and ocean. The human sub-system is formed by communities and their economic, social, and cultural activities. One single change can affect the whole system. The physical, chemical, biological, and social components of the Arctic system are interrelated, and, therefore, a holistic perspective is needed to understand and quantify their connections and predict future system changes (Roberts et al., 2010, p. 3). This highly synergistic relationship of finely co-adapted species requires, from diverse stakeholders, an enhanced understanding and awareness of its functioning and balance. The Arctic environmental and social systems with their variations, impacts, adaptability, and multicultural complexities inspire an impetus for practical improvement in the sense of responsibility, communication, and learning processes.

The critical point continues to be the fragmented, unpractical, and dissonant discussions on crucial human problems. Indigenous and non-indigenous people in the academy have realized the limitations of a mono-cultural education system (Arruda, 2019). Indigenous educators and students have realized the importance of learning about science (chemistry, biology, physics, energy, and scientific theories) while non-indigenous academics started valuing Traditional Knowledge in science and education in adaptive initiatives to enhance awareness. These rich knowledge encounters between indigenous methodology and Western science have the capacity to stimulate and intensify a range of important contemporary conversations on transitional fundamental issues and create convergence about very polarized contemporary debates. This process would be benefited by rethinking the current nature and purpose of our information and knowledge system(s) as well as the alignments of different leadership systems in world-leading economies with various processes of transition currently facing societies.

The new ways of seeing knowledge and dealing with knowledge creation, communication, and dissemination will certainly give space to unheard voices (Arruda, 2019) and will contribute to convergence in knowledge fields in urgent need of rapid consensus. For this to happen, a new approach needs to be discussed, designed, and shared by different audiences/stakeholders from different perspectives and with the potential to enrich the debate of individual and collective responsibility and promote the progress we need to experience in academy, policy-making, and our daily praxis. Educators and universities can have a decisive role in transforming individuals and society.

The first step to form global citizens capable of tackling the challenges of the 21st century will involve expanding consensus and convergence in multicultural dialogue and knowledge systems to promote socio-intercultural understanding (Arruda, 2019) and consequently individual and collective social responsibility. The Arctic presents a natural stage for the transmission of intercultural knowledge and information. Contemporary information technology has been revealed to be an important facilitator of multicultural engagement. Indigenous communities are no longer understood in solely geographical terms but as groups that can use the same technologies to their advantage and to challenge the mainstream representation and narratives (Arruda and Krutkowski, 2017, p. 519). The collaborative effort of using Traditional Knowledge and scientific knowledge can certainly reduce vulnerability and strengthen Arctic communities' resilience and support capacity to implement, in the field, stronger co-management adaptive strategies (Arruda, 2018) in response to climate change, green energy production, natural resources extractivism, and sustainability leadership. In this sense, Education for Arctic Citizenship (Arruda, 2019) goes beyond 'learning as participation' (Vare, 2007) because it represents a change in the structures for learning under the integration of different knowledge systems as per Arruda (2019, p. 95):

> This is learning as sustainability in action. as it is learning that is inculcated in the students' minds as per their practical (active) familiar, geo-experience of creativity and dissemination when living together. It is adaptive open-minded knowledge whose purpose is to effectively impact at local, regional and global arenas. This is the 'education to be a geo-citizen', where the education site respects the 'space' and 'time' of the geographical area at stake, the Arctic.
>
> (Arruda, 2019, p. 95)

A new approach to CSR education should consider not only elements of Arctic Citizenship (Arruda, 2019) but also elements of global citizenship (UNESCO, 2016, p. 6) fundamentally aligned with a self-identification with the whole of humanity (learning to live together), emotional intelligence (compassion), intercultural understanding; ecological literacy (learning to live together sustainably), responsibility, ethics, fairness, and equity. It consists in forming citizens with a human-centred view of development with a sense of belonging to a broader community, beyond national boundaries, emphasizing our common humanity and the interconnectedness between peoples as well as between the local and the global levels. It is a new educational paradigm expressing the ability to understand natural systems through an integrated approach to address environmental problems related to the organization of ecosystems, energy systems, and social systems (UNESCO, 2014; Arruda, 2019). Finally, it implies a higher level of responsibility, ethics, fairness, and equity in relation to attitudes and decision-making (Haigh, 2014, pp. 14, 16). Responsibility is embedded in global citizenship and is expressed in the highest level. It involves living responsibly by understanding that the effects of choices, actions, and decision-making have individual and collective effects at every level of life, family, community, and the whole world (Goleman, 2009). It is based on the universal values of human rights, democracy, non-discrimination, and diversity, and consists of voluntary practices oriented to social justice and global consciousness (UNESCO, 2016, p. 6). A healthy economy, environment, and society can operate within the limits of the planet and enable people to enjoy long and healthy lives, good education, meaningful jobs, physical safety, democratic debate, and a state of well-being.

The future of CSR and the SDGs in the Arctic

Arctic stakeholders have a fundamental role in influencing SDG implementation in the countries in which they operate in terms of aligning discourse, policies, business operations, management, and education to the 2030 Agenda.

In the Arctic of the 21st century, the SDGs have become an integral part of transdisciplinary polar education and research applied to business and management. In its integrated European Union policy for the Arctic, the European Commission emphasized that the EU's actions in the Arctic should contribute to implementing the Agenda 2030 and be in line with the 17 UN SDGs. The SDGs should be regarded as one of the most important aspects guiding the future of polar regions, polar projects, and CSR strategies. The SDGs cannot be implemented independently from each other as they are intertwined multidimensional concepts, which cut across the 2030 Agenda. As a result, coherence is essential to achieve the SDGs, which represent criteria for new governance mechanisms and for forming the global citizens of the 21st century.

Further resources

The Arctic is a developing region and a new frontier for educational innovation. It has been affected by important drivers of change and requires an expanded understanding about the complex dynamics of socio-environmental interactions and adaptation. The concerted participation of stakeholders, information sharing, and local knowledge co-production framed by an inclusive, comprehensive, and systematic sustainability governance and educational policy based on the SDGs and citizens' responsibility will progress the current educational patterns from a mere environmental management approach to a more integrated and embedded version of education oriented to a durable and long-term sustainable development perspective. The focus of this chapter is on higher education but all levels of education would benefit from this alignment with social responsibility, SDGs, Arctic Citizenship (Arruda, 2019), and Global Citizenship, because nurturing and emphasizing these values in early stages of education and at all levels will prepare students to be responsible citizens that live and practice sustainable development as the mainstream approach. This will lead to a time when everything is sustainable and responsible with no distinction between unsustainable and sustainable practices. It will lead to the education of citizens duly equipped to tackle the challenges of the 21st century and generations ahead.

Other additional papers and publications include *Sustainable Energy Education in the Arctic: The Role of Higher Education* (Arruda, 2019) focusing on the nature, structure, and design of the Arctic higher education curriculum to assess levels of sustainable development literacy and the extent to which Arctic higher education courses align with UNESCO Education for Sustainable Development (ESD). This is a discussion of great interest to multicultural students, scholars, and educators with application to higher education across the Arctic and non-Arctic nations. Another relevant resource for developing an ample sense of responsibility and citizenship is *Ecological Intelligence: How Knowing the Hidden Impacts of What We Buy Can Change Everything* (Goleman, 2009), drawing attention to the crucial aspect of consumerism and the need for an expanded understanding in an unselfish and more intelligent attitude about what we buy and use daily and the impacts the manufacturing of these products have on other people's environment and livelihoods. Finally, *Schools in Action: Global Citizens for Sustainable* Development (UNESCO, 2016) is a relevant reading for those interested in changing structures for learning and creating an education environment of responsibility and sustainability in action.

Case study

The discussion in this chapter presents elements of novelty and there is no specific case study that could show an ideal level of advancement in responsibility nor in sustainable responsibility education. However, there are important initiatives underway across the Arctic representing attempts at changing structures for learning in order to enhance responsibility and sustainability in education. One of these initiatives is a comprehensive education review (contents, pedagogies, and curriculum) carried out by the Greenlandic government and the department of education since 2015 aiming at enhancing understanding of the socio-environmental impacts of large-scale natural resources projects (mining, oil and gas, infrastructure, etc.) as well as enhancing the sense of responsibility and sustainability at all educational levels. This initiative is also an attempt to build capacity to support a national education reform taking into consideration Greenlandic geographical complexities where most settlements present a very different reality from towns like Nuuk or Ilulissat. This initiative has a clear commitment to ensuring the delivery of the United Nation's framework of Sustainable Development Goals (SDGs) (United Nations General Assembly, 2015, p. 14) through education. Historical and current challenges have triggered an articulated process of cooperation among national and international stakeholders (scientific, academic, corporative) to allow the development of aligned scientific study programmes and the expansion of physical capabilities as laboratories, research institutes, and business incubators that have been strengthening multidisciplinary research. However, it is crucial to monitor the process and communicate effectively the new achievements made.

Questions for discussion

1. What is the role of education in the CSR of the 21st century?
2. What is the importance of education and sustainable development in a changing world?
3. What is the importance of universities in CSR education?
4. Why should CSR education consider elements of global citizenship?
5. To what extent are CSR and sustainability present in Arctic higher education governance and curriculum?
6. Why is global citizenship education relevant in the 21st century?

Chapter summary

Chapter 10 has highlighted the fundamental role of education for people and planet according to a new vision of education to promote social responsibility and empower citizens. It also drew attention to the necessity of a concerted participation of stakeholders in promoting accessible communication, information sharing, and knowledge co-production. This is framed by a systematic sustainability governance model and educational policy based on SDGs and a sense of citizen's responsibility in moving towards a more integrated and embedded sustainable development perspective.

References

AHDR-II (2014). *Arctic Human Development Report II*. Copenhagen, Denmark: Norden.

Arruda, G.M. (2015). Arctic governance regime: The last frontier for hydrocarbons exploitation. *International Journal of Law and Management*, 57(5), 498–521.

Arruda, G.M. (2014). Global governance, health systems and oil and gas exploration. *International Journal of Law and Management*, 56(6), 495–508. https://doi.org/10.1108/IJLMA-06-2014-0040

Arruda, G.M. (2018). *Renewable Energy for the Arctic: New Perspectives*. Abingdon, UK: Routledge.

Arruda, G.M. (2019). *Sustainable Energy Education in the Arctic: The Role of Higher Education* (1st ed.). Abingdon, UK: Routledge.

Arruda, G.M. and Krutkowski, S. (2016). Arctic governance, indigenous knowledge, science and technology in times of climate change: Self-realization, recognition, representativeness. *Journal of Enterprising Communities: People and Places in the Global Economy*, 11(4), 514–528.

Arruda, G.M. and Krutkowski, S. (2017). Social impacts of climate change and resource development in the Arctic: Implications for Arctic governance. *Journal of Enterprising Communities: People and Places in the Global Economy*, 11(2), 277–288. https://doi.org/10.1108/JEC-08-2015-0040

Battiste, M. and Hernderson, J. (2000). *Sa'ke'j Youngblood: Protecting Indigenous Knowledge and Heritage*. Saskatoon, SK: Purich Publishing.

Berkes, F. and Jolly, D. (2002). Adapting to climate change: Social-ecological resilience in a Canadian Western Arctic community. *Conservation Ecology*, 5(2), 18.

Bitz, C., Blockley, E., Kauler, F., Petty, A., Massonet, F., Arruda, G. M., Sun, N., and Druckenmiller, M. (2016). *Post-Season Report Publishing*. Washington, D.C.: Accessed on 7 January 2020 at www.arcus.org/sipn/sea-ice-outlook/2016/post-season

Boulton, J., Bowman, C., and Allen, P. (2015). *Embracing Complexity: Strategic Perspectives for an Age of Turbulence*. Oxford, UK: Oxford University Press.

Chandler, D. and Werther, W.B. (2013). *Strategic Corporate Social Responsibility: Stakeholders, Globalization, and Sustainable Value Creation* (3rd ed.). Thousand Oaks, CA: SAGE Publications.

Eira, I.M. (2012). *The Silent Language of Snow: Sami Traditional Knowledge of Snow in a Time of Climate Change*. Dissertation for the degree of Philosophiae Doctor, Faculty of Humanities, University of Tromso.

Franz-Balsen, A. and Heinrichs, H. (2007). Managing sustainability communication on campus: Experiences from Lüneburg. *International Journal of Sustainability in Higher Education*, 8(4), 431–445. https://doi.org/10.1108/14676370710823591

Goleman, D. (2009). *Ecological Intelligence: How Knowing the Hidden Impacts of What We Buy Can Change Everything*. New York: Broadway Books.

Gustafson, A. (2016). Cumulative advantage in sustainability communication. *Science Communication*, 38(6), 800–811.

Haigh, M. (2014). From internationalisation to education for global citizenship: A multi-layered history. *Higher Education Quarterly*, 68(1), 6–27.

Kelly, A.V. (2009). *The Curriculum: Theory and Practice* (6th ed.). London: Sage.

Marsh, C.J. (2009). *Key Concepts for Understanding Curriculum* (4th ed.). Abingdon, UK: Routledge.

Peters, M.A., Britton, A., and Blee, H. (2008). *Global Citizenship Education: Philosophy, Theory and Pedagogy*. Rotterdam, the Netherlands: Sense Publishers.

Pittman, J. (2017). Exploring global citizenship theories to advance educational, social, economic and environmental justice. *Journal of Tourism & Hospitality*, 6, 326.

Poppel, B. (2015). *SLiCA: Arctic Living Conditions – Living Conditions and Quality of Life among Inuit, Sami and Indigenous Peoples of Chukotka and the Kola Peninsula*. Copenhagen, Denmark: Nordic Council of Ministers, p. 67.

Radcliffe, S. (2006). *Culture and Development in a Globalizing World: Geographies, Actors, and Paradigms*. Abingdon, UK: Routledge, p. 237.

Roberts, A., Cherry, J., Scott, E., Hinzman, L., and Walsh, J. (2010). Proposed development of a Community Arctic System Model for understanding environmental complexity. *State of the Arctic: Approaches to Integrated Studies of the Arctic System*. Miami, FL: International Arctic Research Center.

Tilbury, D. (2011) *Higher Education in the World 4: Higher Education's Commitment to Sustainability*. Hampshire, UK: Palgrave and Global Universities Network for Innovation (GUNI), p. 3

UN General Assembly. (2015) Transforming our world: The 2030 Agenda for Sustainable Development, 21 October 2015, A/RES/70/1. Accessed on 5 January 2019 from www.refworld.org/docid/57b6e3e44.html

United Nations. (n.d.). Sustainable Development Goals. Accessed on 16 March 2020 from https://sustainabledevelopment.un.org/?menu=1300

United Nations Educational, Scientific and Cultural Organization (UNESCO). (2014) *Education Strategy 2014–2021*. Paris: UNESCO, p. 46. Accessed on 2 December 2018 from www.natcom.gov.jo/sites/default/files/231288e.pdf

United Nations Educational, Scientific and Cultural Organization (UNESCO) (2015) *Global Citizenship Education: Topics and Learning Objectives*. Paris: UNESCO. Accessed on 28 November 2018 from https://unesdoc.unesco.org/ark:/48223/pf0000232993

United Nations Educational, Scientific and Cultural Organization (UNESCO). (2016). *Schools in Action. Global Citizens for Sustainable Development. A Guide for Students*. Paris: UNESCO. Accessed on 25 November 2018 from www.unesco-sole.si/doc/teme/sdg-guide-for-teachers.pdf

van Voorst, R.S. (2009) 'I work all the time – he just waits for the animals to come back': Social impacts of climate changes: A Greenlandic case study. *Jàmbá. Journal of Disaster Risk Studies*, 2(3): 235–252.

Vare, P. (2007) From practice to theory: Participation as learning in the context of sustainable development projects. In A.D. Reid, B.B. Jensen, J. Nikel, and V. Simovska

(Eds.), *Participation and Learning: Perspectives on Education and the Environment, Health and Sustainability*. Dordrecht, the Netherlands: Springer Press.

Willox, A.C., Harper, S.L., Ford, J.D., Edge, V.L., Landman, K., Houle, K., Blake, S., and Wolfrey, C. (2013) Climate change and mental health: An exploratory case study from Rigolet, Nunatsiavut, Canada. *Climatic Change*, 121(2), 255–270. http://dx.doi.org/10.1007/s10584-013-0875-4

Index

Note: Page numbers in **bold** refer to tables, and in *italics* refer to figures.

For Product Safety Concerns and Information please contact our EU representative GPSR@taylorandfrancis.com
Taylor & Francis Verlag GmbH, Kaufingerstraße 24, 80331 München, Germany

www.ingramcontent.com/pod-product-compliance
Lightning Source LLC
LaVergne TN
LVHW021125300926
849160LV00010B/864
* 9 7 8 0 3 6 7 5 2 9 1 8 5 *